Springer Series on Naval Architecture, Marine Engineering, Shipbuilding and Shipping

Volume 7

Series editor

Nikolas I. Xiros, University of New Orleans, New Orleans, LA, USA

The Naval Architecture, Marine Engineering, Shipbuilding and Shipping (NAMESS) series publishes state-of-art research and applications in the fields of design, construction, maintenance and operation of marine vessels and structures. The series publishes monographs, edited books, as well as selected Ph.D. theses and conference proceedings focusing on all theoretical and technical aspects of naval architecture (including naval hydrodynamics, ship design, shipbuilding, shipyards, traditional and non-motorized vessels), marine engineering (including ship propulsion, electric power shipboard, ancillary machinery, marine engines and gas turbines, control systems, unmanned surface and underwater marine vehicles) and shipping (including transport logistics, route-planning as well as legislative and economical aspects).

More information about this series at http://www.springer.com/series/10523

Byung Suk Lee

Hydrostatics and Stability of Marine Vehicles

Theory and Practice

 Springer

Byung Suk Lee
Glasgow, Lanarkshire, UK

ISSN 2194-8445   ISSN 2194-8453   (electronic)
Springer Series on Naval Architecture, Marine Engineering, Shipbuilding and Shipping
ISBN 978-981-13-4800-6   ISBN 978-981-13-2682-0   (eBook)
https://doi.org/10.1007/978-981-13-2682-0

This Springer imprint is published by the registered company Springer Nature Singapore Pte Ltd.
The registered company address is: 152 Beach Road, #21-01/04 Gateway East, Singapore 189721,
Singapore

To my family

Preamble

The word 'hydrostatics' is made up of two parts: 'hydro', which means water, and 'statics', which means characteristics or study of whatever it is that is stationary. It is quite clear, therefore, that by hydrostatics we are talking about the properties of stationary water. A ship normally floats in water, and for the water to be stationary, it is essential that the ship does not move. One can see then that 'hydrostatics' deals with physical systems in static equilibrium. Hydrostatics in naval architecture is concerned with forces, moments and all other properties of stationary ships.

This obviously implies that there is 'hydrodynamics' as well. This sounds really interesting, but anything dynamic is quite a lot more complicated to study and understand than those which are static. In fact, it is not too difficult to see that statics is a special case of dynamics, where the motions happen to be zero.

Let us now go back to still water and consider hydrostatics. Remember that the water is not moving. This means there is no wave and no current, and no wind either, since wind will cause waves and make the ship move, which in turn will disturb water. This cannot represent the reality. We all know that even 'flat calm' mirror-like water surface has ripples on it. Why then are we talking about hydrostatics in naval architecture? An immediate, and not altogether flippant, answer is that it is simpler that way! We can understand quite a lot about water if it does not move.

Actually, there is a slightly more serious answer to it than that. If the motions are not too large, we can treat them as small symmetrical disturbances about a 'mean' or 'static' state. Therefore, by assuming no motion, we are in effect dealing with the mean state. That is not science, but it is quite good engineering. Statics is crucial in understanding dynamics too. Indeed, a mathematical equation describing a cyclical motion has a static element in its restoration term. Take, for example, a simple differential equation representing a damped spring-mass system. Make the acceleration and velocity zero, and what is left is the static term.

A ship designer wants to know quite a few things about the ship being designed. Some of the main ones within the realms of statics can be summarised as follows:

- First of all, will it float?
- How many people, how many cars and lorries, how much volume and weight of cargo can it carry?
- Will it stay upright? (or more importantly, under what condition will it capsize?)
- Will it be strong enough to carry all this cargo and not break up?

This book addresses some of these points and how they can be answered, except the last one which constitutes a separate subject of structural strength.

I will be delighted if the book can help the readers understand the subject of hydrostatics and stability of ships and make the task of learning this important subject a little easier and enjoyable.

Sukie Lee

Acknowledgements

This book would have been impossible without my students. It was originally written as a lecture note to help them study. I am grateful for their complaints about its shortcomings and mistakes over many years.

I was greatly encouraged and exhorted by my family to complete the book when it looked as though it might require too much effort on my part. I am also grateful to them for going through the text to make the sentences a little more readable. I hasten to add, though, that all the grammatical and typographical errors in the book are my own.

Finally, I am grateful to Springer who have agreed to include this little book in their Springer Series on Naval Architecture, Marine Engineering, Shipbuilding and Shipping.

Contents

About the Author

Dr. Byung Suk Lee studied naval architecture at Seoul National University and ship production at the University of Strathclyde in Glasgow. On completion of his Ph.D. at Strathclyde, he joined the academic staff of the Department where he taught ship hydrostatics and stability for over 35 years. During this time, he tried to teach subjects other than hydrostatics. For example, he initiated the introduction of maritime transport and economics into the curriculum among other things and took particular pleasure in trying to teach this intractable subject to the naval architecture students. He also undertook a variety of research projects ranging from stability of ships, hydrodynamics and computer-aided design to container ship operations and safety of LNG-fuelled ships. He has now retired from teaching, although he retains an interest in helping Ph.D. students.

He used to enjoy playing the guitar badly, but his arthritic fingers no longer allow this delight, but he is glad that he can put more energy into growing delicious organic tomatoes. He has recently discovered the joys of making things out of clay, and consequently, the house is beginning to be filled with misshapen pots.

Chapter 1
Basic Ship Geometry

1.1 Introduction

It is superfluous to say that the first main property of all surface ships is that they float on the water whilst carrying payloads. Consequently the main hull of a ship is basically a watertight box. The primary function of most ships is transport in which speed and fuel consumption rate are the two most important factors determining the efficiency. This means that the ships' external forms must be made in such a way that they experience the least resistance while moving in the water while carrying as much load as possible. As a consequence, ships used in transport take on the long and relatively lean shapes that we are all familiar with today. Furthermore, the hull surface is made hydrodynamically 'smooth', so that ship surfaces are complex in shape although often aesthetically pleasing. The shape is so complex that it is virtually impossible to describe it accurately in words alone. This has given rise to two-dimensional representations of the complex hull forms, known as lines plan, and this will be described later. One redeeming feature of the complexity is that most ships have one plane of symmetry, the *centreplane*, giving port-and-starboard symmetry. Some vessels used in offshore industry may even have two or more planes of symmetry (e.g. semi-submersibles). Incidentally, *port* refers to the left hand side of the ship when viewed forward from the stern. *Starboard* is the right hand side.

Since the length of most ships is greater than the breadth, it is natural that the direction parallel to the length is referred to as *longitudinal* direction, as in, for example, 'longitudinal stiffeners' (or simply 'longitudinals'). The athwartship direction is referred to as *transverse* direction, as in 'transverse girders'.

(a) **Major Parts of a Ship**

As stated, a ship is essentially a box, usually water-tight, designed to float and to propel itself carrying passengers and/or cargo. The main body of a ship is known as the hull. The hull is enclosed by the sideshell and the bottom, and, on the top, by the upper (or main) deck, if it is not an open ship. Some structures are built above

© Springer Nature Singapore Pte Ltd. 2019 1
B. S. Lee, *Hydrostatics and Stability of Marine Vehicles*, Springer Series on Naval Architecture, Marine Engineering, Shipbuilding and Shipping 7,
https://doi.org/10.1007/978-981-13-2682-0_1

the deck, used mainly for accommodation, recreation, stores and otherwise running the ship, and these are called superstructures. The spaces within the hull for carrying cargo are called holds, and the openings in the deck to allow access to them are called hatches. Hatches usually have lips around them, called hatch coamings, to stiffen the local structure and to accommodate hatch covers. On ships carrying liquid or gaseous cargo, the holds are called tanks and the cargo is handled through piping systems rather than hatches. Some ships, such as container ships, carry cargo above deck as well.

In most large merchant ships the cross section shape is constant for much of the midship region and this part is known as the parallel middle body. The bow and stern regions are shaped for good hydrodynamic performance in forward motion, and therefore, there is much longitudinal variation in the cross section shape. The forward part is known as 'entry' while the stern part is called 'run'. This point will be discussed later when lines plan is described.

(b) Stations

Before the advent of computers and techniques of defining hull forms mathematically communication of hull forms had to rely on orthogonal section shapes usually at discrete points along the three axes. In particular, the transverse planes at various points along the longitudinal direction are known as stations. It is customary to divide the ship with 11, 21, 31 or 41 evenly spaced stations for form definition, although many more stations are used for ship production when the section shapes at the locations of all the transverse frames have to be defined. Where the hull surface is highly curved, intermediate stations (half or quarter stations) are often used. The foremost station is called forward (or fore) perpendicular (FP) and the aftermost station is called aft perpendicular (AP). Usually the FP is placed where the design waterline (DWL) intersects the stem, whereas the AP is where the DWL intersects the stern or where the rudder stock is. The AP is station 0 and the number is incremented at every whole station going forward, so that the FP is labelled 10, 20, 30 or 40, depending on the number of stations used. Intermediate stations are labelled ½ and ¼: for example, stations 0, ¼, ½, ¾, 1, 1½, 2, 3, and so on. The mid-point between the two perpendiculars is the midship station and is denoted by a symbol ⊗.

These stations were once fundamental in calculating various hydrostatic properties of ships, such as volume displacement and centre of buoyancy among many others, when all calculations were carried out by hand. In this age of computers, however, the hydrostatic calculations do not have to rely on regularly spaced stations. Nevertheless, the old tradition lives on.

(c) Principal Dimensions

Principal dimensions (sometimes called principal particulars) represent the size and major features of the vessel and include the following items (see Figs. 1.1, 1.2 and 1.3):

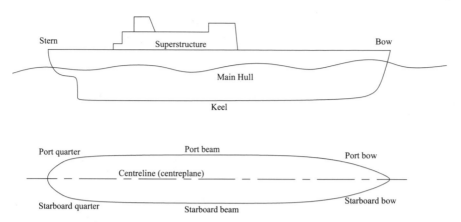

Fig. 1.1 Main regions of a ship

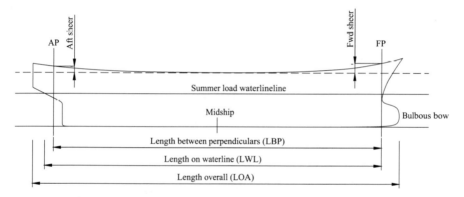

Fig. 1.2 Longitudinal principal dimensions

Length Between Perpendiculars (*LBP*, *L_{BP}* or *L_{PP}*)
It is the distance between the aft perpendicular and the forward perpendicular. This is often the length of the underwater body at the loaded draught, and therefore is used in most hydrostatic, hydrodynamic and other calculations in naval architecture.

Length Overall (*LOA* or *L_{OA}*)
It is the distance from the extreme point at the forward end to a similar point at the aft end. In most ships it exceeds LBP by a considerable amount. The excess includes the overhang of the stern and stem. If there is a bulbous bow extending beyond the end of the bow, LOA is measured to the extreme point of the bulb. LOA determines if a ship can fit in a canal lock, for example.

Length on the Waterline (*LWL* or *L_{WL}*)
It is the longitudinal distance of the waterline at any given draught. For most ships LWL will be different for different draughts.

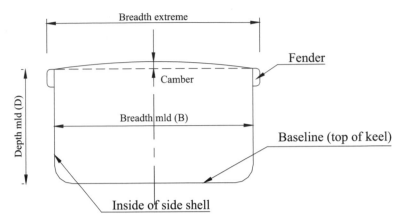

Fig. 1.3 Transverse principal dimensions

Breadth Moulded (B or B_{MLD})

This is the breadth of the ship at its broadest point, measured to the inside surface of the side shell plating. This is the breadth commonly used for most calculations, but is not the greatest breadth of the ship. The breadth extreme is the breadth measured to the outside surface of the shell plating (equal to B_{MLD} plus twice the thickness of shell plating), or any side appendages if any (such as fixed fenders). Breadth extreme will, therefore, determine the width of a waterway that the ship can pass through.

Depth Moulded (D or D_{MLD})

It is the vertical distance measured amidships from the baseline (top of the keel plating) to the underside of the deck plating at side. Note that the depth at centreline is usually greater than that at side, and the difference is known as *camber*.

Draught Moulded (T)

This is the vertical distance from the moulded baseline to the waterline at which the ship is floating. The extreme draught, of course, is the vertical distance from the underside of the keel plating to the waterline. The draught moulded is used for most calculations, but draught extreme is important in determining the minimum depth of water the ship can sail in. It is important to give sufficient clearance because of a phenomenon known as *squatting*.

(d) Weights and Volumes

When a ship is completed and ready for sea, but as yet empty other than just enough fuel and lubricating oil and water to operate the engine, it is known as lightship and its mass *lightweight*. Whatever mass additional to the lightweight that the ship may have on board at any time is called *deadweight* (DWT). The deadweight will include potable water, sludge, fuel and lube oil, stores, provisions, crew and their gear and ballast as well as the *payload*. The sum of the lightweight and the deadweight is equal to the mass of the ship, or *mass displacement* (Δ). This is the same as the mass of the water displaced by the ship in calm water, as discussed in more detail in Chap. 2.

The volume of the water displaced (i.e. volume of the underwater part of the hull) is known as volume displacement (∇). It is obvious, therefore, that

$$\Delta = \nabla \times \rho$$

where ρ is the mass density of water.

The maximum deadweight that can be put on board is limited by the *freeboard*, strength and the seaworthiness of the ship, and is controlled by law (Load Line Regulations). This limitation is shown clearly by markings of the maximum allowable draught on the side of the ship (usually amidships) for a variety of different conditions. These markings are known as 'Plimsoll lines'. The carrying capacity of many cargo ships is represented by their deadweight capacity.

(e) Other Features

There are some items which may be considered minor but sometimes are very important for certain characteristics. These include the following:

Sheer

This is the progressive rise of the deck as one moves from the midship towards the ends. This feature is used to provide as much spare buoyancy as possible at the narrow points of the ship, and also to protect the most vulnerable parts of the ship to some extent. A high bow also reduces the water spray coming on deck while steaming in rough seas. Small fishing boats often show high degree of sheer, allowing the working platform at the side to be as near to water as possible.

In older ships the deck-at-side line was usually a parabola and the sheer is quoted as its value at the forward and aft perpendiculars (see Fig. 1.2). The sheer forward was usually twice as much as the sheer aft. In modern merchant ships, however, the deck at side line can either be flat with zero sheer over some distance on either side of the midship and then rise as a straight line towards the ends, or be completely flat with no sheer over the entire length.

Rise of Floor

Some ships have a flat bottom (usually large ships), but others may have a bottom rising from the centreline or a little off it. The height of the intersection of a line drawn along this rising floor and the vertical line at the moulded breadth is called the rise of floor. High speed vessels often have a high rise of floor (deep V-section) at the bow region which becomes smaller (shallow V-section or U-section) towards the stern.

Freeboard
This is the vertical distance from the waterline to the deck at side. Usually it will be a minimum amidships and will increase towards the ends (due to sheer). Obviously this is related to the maximum allowable load line, and the minimum permitted freeboard is governed by the Load Line Regulations.

1.2   Form Definitions

(a)   Lines Plan

Except for a very few exceptional cases, nearly all ships have hull surfaces which have three-dimensional curvature. This means that the hulls cannot be represented by simple common geometric elements. Indeed many techniques had to be devised in the last few decades to represent the hull surfaces in mathematical/numerical forms so as to accommodate computerisation. However, the traditional method of representing the hull form relies on section shapes in planes parallel to the three major orthogonal planes. The drawing containing these sectional shapes is known as a *lines plan* (see an example in Fig. 1.4) and consists of three plans as follows:

Half Breadth Plan or **Waterlines Plan** (intersections of horizontal planes with ship's moulded surface)
It shows a collection of waterlines at various draughts. Each line represents the shape of a *waterplane*. The waterplanes appear as straight lines when viewed from the front or side, and consequently they are more frequently referred to as waterlines. The load waterplane (LWP), or load waterline (LWL), represents the waterplane at which the ship is designed to float in fully loaded condition. A design waterline (DWL) is the waterplane for which the ship is designed. This does not mean the ship cannot operate at any other draught, but that the designer tried to optimise the performance of the ship for that draught. This, as you can imagine, is often the same as the LWL. The plan is known as 'half breadth plan' because only one half of the waterlines needs to be shown because of the port-and-starboard symmetry of all ships.

In order to ensure fair hull form some section shapes of longitudinal planes other than the two orthogonal planes are also drawn below the centreline of the half-breadth plan. These are known as diagonals, and are of particular significance for sailing yacht forms, as they often sail with a static heeling angle, and, therefore, 'fairness' of the lines when heeled is crucial for their performance.

Body Plan (intersections of transverse vertical planes with moulded hull surface)
It is a collection of transverse sections at regular, designated stations along the ship's length. For normal purposes, only one half of the section shape is shown. The forward half of the ship is shown on the right hand side of the centreline and the aft half is shown on the left hand side. The sections are shown for all the numbered stations as described above. For example, a basic design lines plan may have stations 0, ¼, ½, ¾, 1, 1½, 2, 3, 4, 5, 6, 7, 8, 8½, 9, 9¼, 9½, 9¾, 10. Normally station 0 corresponds

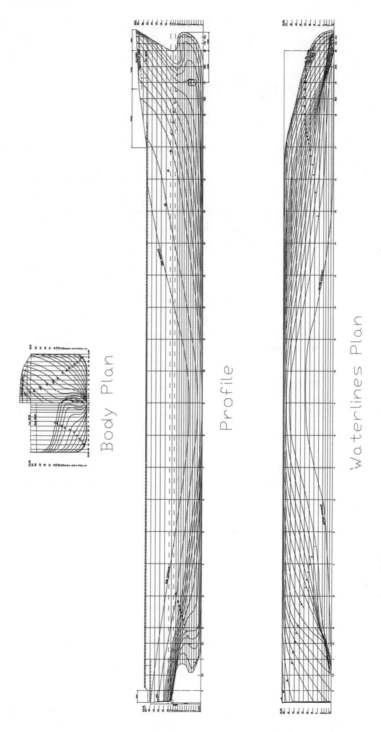

Fig. 1.4 An example of lines plan (by courtesy of Dr. JY Kang)

to the AP, and station 10 the FP, although some people have used 0 for FP and 10 for AP. Sometimes extra stations are placed outside these two perpendiculars. Station 5 corresponds to midship. Body plans used for ship production are drawn for the locations where transverse frames and girders are placed. The appropriate frames are made from steel members bent, and girders are made of steel plates cut to the section shapes. Many merchant ships have a long parallel middle body and the sections within this are identical.

Sheer Plan or **Profile** (intersections of longitudinal vertical planes with the moulded hull surface)
The sheer plan shows longitudinal sections of the hull on the planes parallel to the centreplane at a number of offsets. The forward lines are known as bowlines, while the aft lines are called buttock lines or buttocks. For monohull vessels the line at 0 offset will show the profile outline of the vessel together with the deck at centre line. The plan can also include a deck at side line.

A lines plan is usually a very long drawing, and to reduce the inconvenience of having to deal with such long drawings, it is sometimes drawn with the forward half superimposed on the aft half, as though the ship is folded in half.

The importance of lines plan has diminished somewhat in recent years due to the fact that computer-aided design is wide-spread in ship and boat design, and the design programs use mathematical methods (e.g. NURBS—non-uniform rational B-spline surface) of representing hull forms. Nevertheless, all design documentation still contains a lines plan.

(b) Table of Offsets

An offset table provides the information contained in the lines plan in a tabular form. Half breadths are arranged with columns for stations and rows for waterlines, or the other way around. In addition, other information, such as sheer, may also be included. A lines plan is nearly always accompanied by the corresponding table of offsets.

(c) Form Coefficients

There are a number of non-dimensional coefficients which are used to indicate the major geometric characteristics and to give an idea of the performance of the vessel. For geometrically similar vessels the form coefficients are identical.

Block Coefficient (C_B)
This is the ratio of the volume of the underwater form (∇) to the volume of a rectangular block whose sides are equal to the immersed length, beam and draught (Fig. 1.5). Thus,

$$C_B = \frac{\nabla}{L_{BP} \times B \times T}$$

Imagine someone whittling away on a rectangular block of wood to make the underwater body of a ship. Then, $(1 - C_B)$ represents the portion of the wood to be whittled

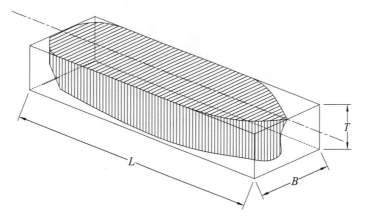

Fig. 1.5 Block coefficient

Fig. 1.6 Midship section
coefficient

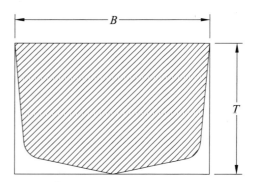

away. The block coefficient, therefore, gives an indication of the fullness of the form
and its value may vary from less than 0.4 for some high speed vessels, yachts and mil-
itary vessels through around 0.60 for passenger ships and aircraft carriers to nearly
0.9 for large crude oil carriers and bulk carriers.

Midship Section Coefficient (C_M)
This is the ratio of the immersed area of the midship section (A_M) to the area of the
enclosing rectangle whose sides are equal to B and T (see Fig. 1.6). Thus,

$$C_M = \frac{A_M}{B \times T}$$

This coefficient expresses the fullness of the midship section and its value is
usually around 0.75–0.98. For extreme forms it may be as low as 0.65 or as high as
1.0.

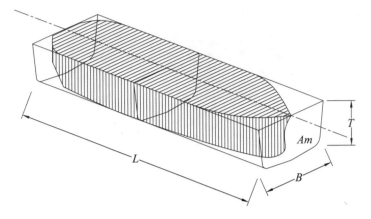

Fig. 1.7 (Longitudinal) prismatic coefficient

A similar definition of the maximum section coefficient (C_X) exists, but since the midship section is nearly always the maximum section, it is usually identical to the midship section coefficient.

Prismatic Coefficient (C_P)
This is the ratio of the volume of the underwater form to the volume of a prism having a constant cross section of the immersed midship section and length equal to L_{BP} (Fig. 1.7). Thus,

$$C_P = \frac{\nabla}{A_M \times L_{BP}}$$

This coefficient is sometimes called longitudinal prismatic coefficient to distinguish it from the vertical prismatic coefficient (see below). It is a very important parameter determining the forward motion resistance of the vessel. As can be expected, the ships with lesser resistance will be finer in form and therefore will have smaller C_P. Its value may range from about 0.55–0.80.

The prismatic coefficients of the forward half and the aft half of the hull are sometimes used separately to help determine the form of the *entry* and *run*.

Vertical Prismatic Coefficient (C_{VP})
This is the ratio of the volume of the underwater form to the volume of a cylinder having a depth equal to the vessel's mean draught (T) and a constant horizontal cross section identical to the waterplane at that draft. Thus,

$$C_{VP} = \frac{\nabla}{A_W \times T}$$

where A_W is the waterplane area.

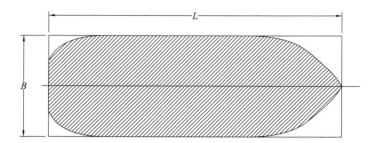

Fig. 1.8 Waterplane area coefficient

Table 1.1 Examples of principal dimensions

	Tug	Ferry	Cargo	Cargo	Passenger	Tanker
L_{BP} (m)	32.1	95.3	113.0	213.0	362.0	270.0
B (m)	8.60	25.1	18.0	32.1	47.0	48.2
T (m)	3.50	2.45	7.5	11.9	9.33	16.0
Δ (tonnes)	390	3560	9954	53,217		
A_M (m^2)	20.50	60.5	133.5			
TPC (tonnes/cm)	1.99	20.1		50.2		
C_B					0.615	0.716
C_P				0.645		
C_{VP}						
C_W			0.80		0.745	0.88
C_M					0.974	0.99

Note TPC will be discussed in Chap. 2

Waterplane Area Coefficient (C_W)
This is the ratio of the waterplane area to its circumscribing rectangle (Fig. 1.8). Thus,

$$C_W = \frac{A_W}{L_{WL} \times B}$$

Unless otherwise stated, C_W is for the load waterline of the vessel, at which L_{WL} is often the same as L_{BP}.

Table 1.1 gives sample principal dimensions and/or some form coefficients. Perhaps you would like to calculate the items left blank and compare them between various ship types.

Key Points

Block coefficient $C_B = \frac{\nabla}{L_{BP} \times B \times T}$

Midship section area coefficient $C_M = \frac{A_M}{B \times T}$

Prismatic coefficient $C_P = \frac{\nabla}{A_M \times L_{BP}}$

Waterplane area coefficient $C_W = \frac{A_W}{L_{WL} \times B}$

Chapter 2
Flotation

2.1 Archimedes' Principle

Most people have heard, usually at school, the funny but exhilarating story of Archimedes running through the street of ancient Syracuse stark naked shouting 'Eureka!' Well he might!

We all experience lightening of our body when we get into a bath and therefore it was not this phenomenon itself that he discovered. The significance of his discovery is in quantifying that experience precisely, thereby enabling the application of this principle in 'engineering'. Although we now take this momentous discovery for granted, it is so fundamental to naval architecture that it is necessary to examine his principle with the help of more modern understanding.

Archimedes' Principle states that when a body is immersed in a quiescent fluid, the body experiences an upward force equal and opposite to the weight of the fluid displaced by it. His initial interest in this principle is that it enabled one to obtain the volume of any irregular shaped object. However, we are more interested in this upward force here, which is called buoyancy force. We now know that the buoyancy is the resultant of all the normal pressures exerted by the fluid on the entire immersed surface of the body. This pressure is known as hydrostatic pressure and is proportional to the depth of immersion, or

$$P = \rho g h$$

where

P hydrostatic pressure at any point on the immersed body surface
ρ mass density of the fluid
g gravitational acceleration (≈ 9.806 m/s^2)
h depth of immersion of the point.

© Springer Nature Singapore Pte Ltd. 2019
B. S. Lee, *Hydrostatics and Stability of Marine Vehicles*, Springer Series on Naval Architecture, Marine Engineering, Shipbuilding and Shipping 7,
https://doi.org/10.1007/978-981-13-2682-0_2

14

For an arbitrary body floating in static equilibrium in a still fluid shown in Fig. 2.1, the vertical force (or the buoyancy force) B is the sum of the vertical component of the surface pressure, P, multiplied by the area that it is applied to, ds, or

$$B = \int_S P \cos\theta \cdot ds = \int_S \rho g h \cdot ds \cdot \cos\theta$$

It is evident that $h \cdot ds \cdot \cos\theta$ is in fact the immersed volume of the elemental column which presents the surface ds to the fluid at an angle θ at its lower end. Therefore, $\int_S h \cdot ds \cdot \cos\theta$ is the total immersed volume of the body. Let the volume be V, then

$$B = \rho g V$$

This is true, provided ρg is constant, which is practically true for incompressible fluid and for the kind of depth we are concerned with. Since $\rho g V$ is the weight of the fluid of the volume displaced by the immersed body, the crucial Archimedes' discovery is thus proved.

One can also prove easily that the sum of the horizontal component of the pressure is zero, and therefore the body does not experience any pure horizontal force when simply immersed.

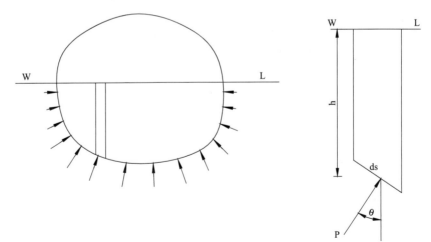

(a) Hydrostatic pressure on the immersed surface   (b) An elemental volume

Fig. 2.1 Mechanism of buoyancy force

> **Key Point**
>
> The buoyancy force acting on a body either fully or partly submerged in a fluid is the same as the weight of the fluid displaced by the body, or
>
> $$B = \rho g V$$

2.2 Volume and Mass Displacement

The volume of the underwater part of a ship's body is known as *volume displacement* and is represented by the symbol ∇ (pronounced 'nabla'). The normal unit for volume displacement is m^3. The mass of the water displaced is called *mass displacement* and is represented by the Greek Δ (pronounced 'delta'). Thus, $\Delta = \rho \nabla$. The unit of mass displacement is usually tonnes (the mass of 1 L of fresh water is 1 kg, and that of 1 m^3 of fresh water is 1 tonne which is equal to 1000 kg). Note that Δ is the same as the *mass* of the ship and $g\Delta$, sometimes called force displacement, will be the same as the *weight* of the ship. Indeed one hardly ever sees any special symbol for the ship's weight, since it can be represented by the mass displacement. It is customary that a ship's displacement refers to its mass displacement. The unit of force displacement is N, but since mass is in tonnes, it is more convenient to use kN (weight of 1 tonne is approximately 9.806 kN).

2.3 Static Equilibrium

Although the buoyancy force is distributed all over the *wetted surface*, it is often more convenient to consider that the total buoyancy force acts vertically upward at or from a single point. This is the point about which the moment of the hydrostatic pressure, when summed over the entire wetted surface, becomes zero. This point is called the *centre of buoyancy*. In a similar manner the gravitational force (i.e. the weight of the object) can also be regarded as acting vertically downward from one point: this is the *centre of gravity*.

In general, a state of equilibrium exists in a system if and only if

(a) the sum of all the forces is zero; and
(b) the sum of all the moments is zero.

An object freely floating in still water, therefore, will have just the right volume immersed to provide exactly the same buoyancy force as its weight, and will take an attitude which puts the centre of gravity vertically in line with the centre of buoyancy. Their vertical relative position does not matter, although it has a significant role in the stability of the system. This state is known as static equilibrium in contrast to dynamic equilibrium which is a subject which requires a separate treatment. Of course, if its total volume cannot provide sufficient buoyancy, it will simply sink. If the total volume can produce buoyancy exactly the same as the weight, it will be *neutrally buoyant*, as submarines can be in certain conditions.

If the first of the two equilibrium conditions is not met, then the object will either sink further or rise until that condition is fulfilled. If only the second condition is not met, then the object will rotate itself until the equilibrium is reached. In general, however, these two processes occur simultaneously.

2.4 Curve of Displacement

Unlike objects of simple geometric shapes, the volume displacement of a ship for any given draught is not simple to calculate. These days, this can be done by a computer at a blink of an eye, and thus presents no great difficulty. However, in the days when every single computation had to be done laboriously by hand, it was immensely important to devise methods of instantly reckoning the expected draught for a given loading condition or the mass displacement for a given draught.

The volume displacement calculated for a range of draughts can be plotted against draught, and the curve is smoothed (naval architects are, if nothing else, past masters of generating smooth curves). From this curve, the volume displacement at any mean draught can be readily obtained and vice versa. This curve presupposes that the ship will have a fixed pre-defined longitudinal attitude (trim). Trim, however, is the subject of a later chapter.

2.5 Tonnes per cm Immersion (TPC)

The loading condition of a ship changes often and it is important to keep track of what draught the ship will have at the end of a minor loading/unloading operation. For this the curve of displacement may be used, but for obtaining a quick indication of the change of draught for a given change in mass displacement, a ratio known as tonnes per cm immersion is very useful. The TPC of a ship at any given draught is the mass required to increase the draught by one cm (parallel to the existing waterline, and hence the terms *parallel sinkage* and *parallel rise*).

Consider a vessel floating at a draught T in water of density ρ (by the way, 1.025 or 1.026 tonnes/m^3 is often taken for the normal mass density of sea water). Suppose that its waterplane area is A_W at this draught, and some cargo is now placed on the vessel so that it sinks by 1 cm. This increases the underwater volume by approximately

$$d\nabla = A_W \times 1 \text{ cm}$$
$$= A_W \times 0.01 \text{ (in m}^3\text{, if } A_W \text{ is in m}^2\text{)}$$

Therefore, the increase in mass displacement, or TPC, is

$$d\Delta = d\nabla \times \rho = 0.01\rho A_W$$

Note that this increase in displacement will be identical to the mass of the cargo loaded which caused the parallel sinkage.

Conversely, we can find the parallel sinkage or rise caused when a mass m is loaded or unloaded as follows

$$\text{parallel sinkage} = m/TPC$$

A moment's reflection will show that TPC is in fact proportional to the gradient of the curve of displacement. Note also that in the above discussion we have assumed the layer of the body which is immersed by the increased mass of the object has a constant waterplane area. Strictly speaking, this is not true, but most ships are virtually 'wall-sided', and it is accurate enough for most engineering purposes for small changes in draught. Indeed this type of approximation, simplification and linearisation used to be essential before the advent of computers. Remember that computers began their serious inroads into ship design only in the 1970s!

2.6 Effect of Water Density

Ships travel, and the seas of different areas have different mass density due to changes in salinity and temperature. A more pronounced change in mass density of water is experienced by ships on moving from fresh water into sea water (as will be the case in the area of the Great Lakes and St. Laurence Seaways). There is about 2.5% difference in the density of fresh water and sea water, and obviously this will cause sinkage or rise of the vessel during the transit. In all these cases, only one principle should be remembered: 'the mass displacement of the ship should be the same as the mass of the ship at all times, whatever the density of water.' This necessitates a change in the underwater volume which can only be achieved by altering the draught.

Consider a vessel of displacement 15,000 tonnes floating in sea water of density 1.025 tonnes/m^3 at a draught of 7 m. The volume displacement in sea water is

$$\nabla_S = 15,000/1.025 = 14,634 \text{ m}^3$$

When this ship moves into fresh water of density 1.0 tonne/m^3, its volume displacement has to be

$$\nabla_F = 15,000/1.0 = 15,000 \text{ m}^3$$

The change in volume displacement required is

$$dV = V_F - V_S = 366 \text{ m}^3$$

This increase in volume displacement of 366 m³ can only be achieved by the (parallel) increase of draught. If the waterplane area of the ship is 2000 m² at 7 m draught, then the parallel sinkage is

$$dT = 366/2000 = 0.183 \text{ m} \quad \text{or} \quad 18.3 \text{ cm}$$

Note that in general, the new waterline will not necessarily be parallel to the original waterline, but it is convenient to consider the parallel rise or sinkage first and then the change in trim separately.

Key Points

- The buoyancy force acting on a body either fully or partly submerged in a fluid is the same as the weight of the fluid displaced by the body, or $B = \rho g V$
- $\Delta = \rho V$
- In general, a state of equilibrium exists if and only if $\Sigma F = 0$ and $\Sigma M = 0$.
- $TPC = 0.01 \rho A_W$
- $dT = {}^{d\Delta}/_{100} \times TPC$

Exercises

1. A hollow steel cylinder is 1.83 m in diameter, 7.62 m long and has a mass of 16.26 tonnes. At what draught will it float in sea water ($\rho = 1.025$ tonnes/m³), if it floats so that its longitudinal axis is vertical?
   What will then be the hydrostatic pressure at any point on the bottom? Calculate the pressure from the weight of the cylinder and compare it with the hydrostatic pressure.

2. When a ship, without changing weight, passes from sea water to fresh water, it will sink further in the water, because fresh water is less dense than sea water. Show that the increase in draught, say d (in cm), is given by

$$d = \frac{\Delta(\rho_s - \rho_f)}{\rho_f \cdot TPC}$$

where

Δ displacement of the ship in tonnes
ρ_s density of sea water in tonnes/m^3
ρ_f density of fresh water in tonnes/m^3
TPC   is for sea water.

3. Show that for any ship

$$\frac{C_P}{C_{VP}} = \frac{C_W}{C_M}$$

4. A ship, 122 m long, has a beam of 15.25 m and floats in sea water with an even keel draught of 5.50 m. If the block coefficient, C_B, is 0.695, what is the displacement?
    The immersed midship section area is 82.50 m^2. Calculate the values of C_P and C_M.

5. A ship has the following particulars:
    L$=$128 m, B$=$19.20 m, T$=$8.85 m, C_B $=$0.713, C_M $=$0.945.
    Calculate

    – Moulded (mass) displacement in sea water
    – Area of immersed midship section
    – Prismatic coefficient.

6. A ship of displacement 1747.0 tonnes has a beam-to-draught ratio equal to 3.53 when floating in sea water. If the immersed midship section is 30 m^2, $C_B = 0.537$ and $C_M = 0.834$, calculate the length, beam and draught of the ship.

7. A ship of 100 m$_L$ $\times$ 25 m$_B$ operating in sea water at an even keel draft of 5 m has displacement of 7175 tonnes. Its C_w is 0.8 and C_m is 0.92.

    – Calculate its C_b and C_p.
    – Calculate TPC.
    – What will be the new mean draught if 250 tonnes is unloaded?

8. A ship of conventional hull with the following characteristics is operating in sea water:

LBP = 124.5m
B = 24 m
T = 8.035 m
WPA = 2234.5 m^2
C_b = 0.65
Midship section area = 177.413 m^2

(a) Compute the volume displacement, and hence mass displacement, and the prismatic coefficient.
(b) If the ship sails into a fresh water lake, what is the new draught? Assume that there is no change in trim.
(c) Compute the new block, new midship section area and prismatic coefficients.

Chapter 3
Moments and Centroids

One of the most important concepts in engineering is moments, of various order, of lines, areas, volumes, masses and forces. One can even talk of a moment of a moment. Indeed, the area of a 2-D shape can be called its 0th moment (the lever is brought up to the power of 0). Of these infinite number of moments, the first and second moments are encountered the most often in various disciplines of naval architecture. Here we shall confine our discussion up to the second moments of area, volume and mass.

3.1 Area

The area of an arbitrary two-dimensional (2-D) shape in the xy-plane shown in Fig. 3.1 can be calculated by first considering an infinitesimally small rectangle of width dx and height dy within the shape. The area of this rectangle is $dA = dxdy$. Therefore, the whole area of this shape can be obtained by integrating this elemental area throughout the shape:

$$A = \int_A dA = \int_A dx \cdot dy$$

The symbol $\int_A$ means integration over the whole area A.

© Springer Nature Singapore Pte Ltd. 2019
B. S. Lee, *Hydrostatics and Stability of Marine Vehicles*, Springer Series on Naval Architecture, Marine Engineering, Shipbuilding and Shipping 7,
https://doi.org/10.1007/978-981-13-2682-0_3

3.2 First Moment of Area

The first moment of an infinitesimally small rectangle about a given axis can be obtained by multiplying the area with the distance between that axis and the rectangle (lever). The word 'infinitesimal' is used here to signify that the area is so small that it can be assumed to be a point. The lever from the reference axis to that area is simply the shortest distance from the point to the axis. In a Cartesian co-ordinate system the first moment about y-axis will have the lever parallel to x-axis as shown in Fig. 3.1. The first moment of an area is analogous to the moment of force, often known simply as moment, where the force is multiplied by the lever.

Consider the arbitrary enclosed 2-D shape studied above. The first moment of the infinitesimally small area dA about y-axis can be calculated by multiplying the area dA with the lever to y-axis x:

$$dM_y = x \cdot dA$$

Therefore, the moment of A about y-axis M_y is

$$M_y = \int_A x \cdot dA$$

Similarly

$$M_x = \int_A y \cdot dA$$

Fig. 3.1 An arbitrary 2-D shape

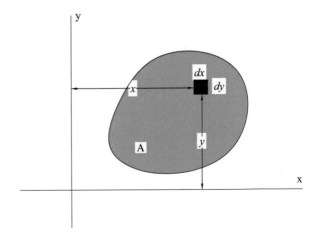

The lever (x or y in this case) can be positive or negative. It follows, therefore, that a first moment of anything can be positive, zero or negative. Indeed it can be seen at once that the first moment of an area about the axis of symmetry (if it exists) is zero.

Furthermore, in Fig. 3.1, M_y is obviously positive, since all the levers are positive; similarly M_x is also positive. Consider now another extreme case where the whole shape lies entirely on the left hand side of the y-axis and well below the x-axis. Then, both moments will be negative, since the levers are always negative. This means that as the y-axis moves from the left hand side of the area to the right, it will pass a point at which M_y is zero—keep the y-axis there. Similarly as the x-axis moves upwards from below the area, it will pass a point at which M_x is zero, and we keep the x-axis there. The intersection of these two axes is known as the ***centroid*** of the area.

The definition of the centroid of an area, therefore, is that it is the point of intersection of any two axes the first moments of the area about which are zero.

Consider an arbitrary 2-D area A and two parallel axes distance h apart, say y and y' with y' on the left of y, as shown in Fig. 3.2. Then the first moments of the area about these axes are

$$M_y = \int_A x \cdot dA \text{ and}$$

$$M_{y'} = \int_A x' \cdot dA = \int_A (x+h)dA = \int_A x \cdot dA + \int_A h \cdot dA$$

$$= M_y + h \int_A dA = M_y + h \cdot A$$

Fig. 3.2 Moment for two parallel axes

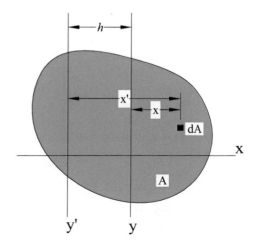

If we now assume that the axis y passes through the centroid, then

$$M_y = 0$$

Therefore

$$M_{y'} = h \cdot A$$

In other words, the first moment of an area can be obtained by multiplying the area with the distance from the centroid to the reference axis. Conversely, the position of the centroid relative to any reference axis can be found by dividing the first moment of the area about that axis by the area. Actually these two principles are applicable to any kind of first moments, and are used in determining centre of gravity, centre of buoyancy, centre of flotation and so on.

Key Points

- The first moment of an area about any axis passing through its centroid is 0.
- The first moment of an area about an arbitrary axis can be obtained by multiplying the area with the distance from the centroid to that axis.
- Conversely, if the first moment of an area about an arbitrary axis is known, the distance from this axis to the centroid can be found by dividing that first moment by the area.

Identifying the centroid of the waterplane for a given draught is very important, as it is required in estimating the longitudinal stability and trim as will be discussed in later chapters. Normal ships have port-and-starboard symmetry. Therefore, it is obvious that the centroid of the waterplane will be on the centreline. It remains to identify the longitudinal location of the centroid, and this can be done for any reference point, but normally it is done for the midship. The longitudinal distance from the midship to the centroid of waterplane is known as LCF (standing for Longitudinal Centre of Flotation), the reason for which will be discussed when we study the longitudinal stability.

Consider a waterplane shown in Fig. 3.3. The origin of the axis system is on the waterline amidships, so that LCF can be estimated relative to the midship directly without any adjustment.

The first moment of the waterplane about y-axis is

$$M_y = 2 \int_{-L_a}^{L_f} xy\,dx$$

Fig. 3.3 A waterplane with the usual axis system

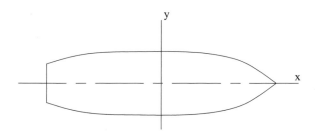

where

L_a and L_f are the length of aft body and fore body respectively,

y is the half-breadth of the waterplane and is a function of x.

The area of the waterplane is

$$A_W = 2 \int_{-L_a}^{L_f} y \, dx$$

Therefore, midship to centroid is

$$LCF = \frac{M_y}{A_W} = \frac{\int_{-L_a}^{L_f} xy \, dx}{\int_{-L_a}^{L_F} y \, dx}$$

Note that, in this notation, a positive LCF indicates the centroid is forward of the midship. Most often longitudinal positions of all kinds are given as 'aft or fwd of midship' rather than as signed numbers. Nevertheless, having a sign convention is invaluable in many calculations as long as it is applied consistently.

Example 3.1 Consider the rectangle $a \times b$ shown in Fig. 3.4.

An infinitesimally thin strip of thickness dy within the rectangle has area $a \cdot dy$. Since the distance of this strip from x-axis is y, its first moment about x-axis is $ay \cdot dy$.

Fig. 3.4 A rectangle

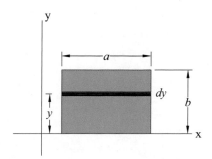

Fig. 3.5 A right-angled
triangle

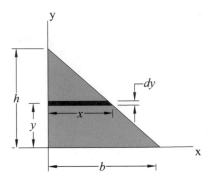

Therefore, the first moment of the whole rectangle about x-axis is

$$M_x = \int_0^b aydy = \left[\frac{1}{2}ay^2\right]_0^b = \frac{1}{2}ab^2$$

Since the area of the rectangle is $A = ab$, the y-coordinate of the centroid is

$$y_c = \frac{M_x}{A} = \frac{1}{2}ab^2 \bigg/ ab = \frac{1}{2}b$$

Try to calculate M_y and x_c in a similar way.

Example 3.2 We now try to find the first moment and thus the position of the centroid
of a right-angled triangle. Consider the triangle shown in Fig. 3.5.

The first moment of an infinitesimally thin horizontal strip of thickness dy about
x-axis is

$$dM_x = ydA = yxdy,$$

x can be expressed in y using the equation representing the hypotenuse as

$$x = b - \frac{b}{h}y$$

The first moment of the triangle, therefore, is

$$M_x = \int_0^h yxdy = \int_0^h \left(by - \frac{b}{h}y^2\right)dy = \left[\frac{b}{2}y^2 - \frac{b}{3h}y^3\right]_0^h = \frac{bh^2}{6}$$

Since the area of the triangle is $A = \frac{bh}{2}$, the y-coordinate of the centroid is

$$y_c = \frac{bh^2}{6} \bigg/ \frac{bh}{2} = \frac{h}{3}$$

Try to calculate the first moment of this triangle about y-axis and the x-coordinate of the centroid yourself in a similar way.

3.3 Second Moment of Area

The second moment of an infinitesimally small area about a given axis is the area multiplied by the distance between the axis and the area (lever) squared. A second moment of area is not as easy to visualise as the first moment, but it is analogous to a second moment of mass, which is also known as mass moment of inertia.

The second moment of an area about x-axis is often represented by I_x or I_{xx}, and it is in fact a moment about x-axis of the first moment of the area about x-axis. Note that one can have a moment about y-axis of the first moment of the area about x-axis, and this is known as 'product of area' and represented by a symbol I_{xy}.

Consider an arbitrary area of Fig. 3.6. In this case

$$I_x = \int_A y^2 dA \text{ and } I_y = \int_A x^2 dA$$

(a) Parallel Axis Theorem

Consider an arbitrary area shown in Fig. 3.7. y-axis passes through the centroid of the area, and an η-axis is parallel to it at a distance l.

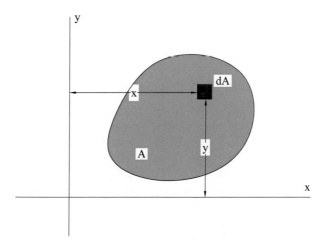

Fig. 3.6 An arbitrary 2D object on x-y plane

Fig. 3.7 Parallel axis
theorem

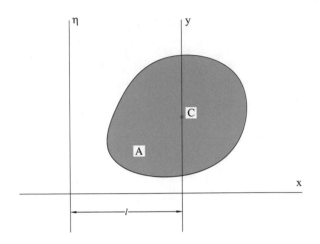

Then

$$I_y = \int_A x^2 dA$$

and

$$I_\eta = \int_A (x+l)^2 dA = \int_A (x^2 + 2xl + l^2) dA = \int_A x^2 dA + \int_A l^2 dA + 2\int_A xl dA$$

Since l is a constant,

$$I_\eta = \int_A x^2 dA + l^2 \int_A dA + 2l \int_A x dA = I_y + Al^2 + 2lM_y$$

where

A is the area of the shape and M_y is first moment of the area about y-axis.

However, since y-axis passes through the centroid, $M_y = 0$. In other words,

$$I_\eta = I_y + Al^2$$

This means that the second moment of an area about any axis is equal to the second moment of that area about an axis parallel to that axis passing through the centroid of the area plus the area multiplied by the distance between the two axes squared. This is known as the *parallel axis theorem*, and it applies equally to second moment of mass and volume as well. It is a very important principle useful in many branches of engineering.

This theorem can be applied directly to any two parallel axes, as long as one of them is a centroidal axis. If neither of the two parallel axes passes through the centroid, this theorem can only be applied by calculating the second moment about the parallel centroidal axis in the first instance.

The second moment of a waterplane area about the transverse axis passing through its centroid (i.e. centre of flotation) is known as the longitudinal second moment of waterplane area, or I_L (because the lever is in the longitudinal direction) and that about the centreline is referred to as the transverse second moment of waterplane area, or I_T (lever is in the transverse direction). We shall deal with practical applications of these items in due course.

Key Points

- The second moment of an area about an axis is the moment of the first moment of the area about that axis.
- Parallel axis theorem: $I_n = I_y + Al^2$
- It is important to remember that one of the two axes must be a centroidal axis, in this case y-axis.
- I about the centroidal axis is always smaller than I about a parallel axis away from the centroid.
- Longitudinal second moment of a waterplane (I_L) is about the transverse axis passing through the longitudinal centre of flotation.
- Transverse second moment of a waterplane (I_T) is about the centreline.

(b) Second Moment of a Rectangle

Consider a rectangle ($b \times h$) shown in Fig. 3.8. We wish to find the second moment of this rectangle about an axis passing through the centroid and parallel to the b-side in the first instance and then calculate the second moment of this rectangle about the h-side.

Fig. 3.8 A rectangle of dimensions $b \times h$

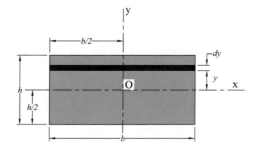

As illustrated in the figure, we put a Cartesian co-ordinate system with the origin at the centroid of the rectangle and the x-axis parallel to the b-side and the y-axis parallel to the h-side. Consider a thin strip of width dy, lengthwise parallel to the x-axis and at a distance y away from the x-axis. Its area is bdy and its second moment about the x-axis is

$$dI_x = y^2 b dy$$

Therefore the second moment of the whole rectangle about the x-axis is

$$I_x = \int_{-h/2}^{h/2} y^2 b dy = \left[\frac{1}{3}y^3 b\right]_{-h/2}^{h/2} = \frac{bh^3}{12}$$

Using this result, we can also calculate the second moment about the b-side. The distance between the x-axis and the b-side is $h/2$. Since x-axis is a centroidal axis, we can apply the parallel axis theorem directly, thus

$$I_b = I_x + bh\left(\frac{h}{2}\right)^2 = \frac{bh^3}{12} + \frac{bh^3}{4} = \frac{bh^3}{3}$$

(c) Second Moment of a Triangle

Consider a right angled triangle with two orthogonal sides of length b and h as shown earlier in Fig. 3.5. We wish to find the second moment of this triangle about the side of length b, and then, using this result, calculate its second moment about an axis parallel to that side, but passing through the centroid of the triangle.

Put the x-axis along the b side, and the y-axis along the h side. Then the hypotenuse can be expressed as

$$y = -\frac{h}{b}x + h \text{ or } x = b - \frac{b}{h}y$$

We can calculate the second moment in two different ways. The first method takes a narrow strip of width dy, lengthwise parallel to and at a distance y from the x-axis. Since its length is x which can be expressed as $b - \frac{b}{h}y$, its area is

$$dA = xdy = \left(b - \frac{b}{h}y\right)dy$$

Its second moment about x-axis is (remembering that dy is infinitesimally small)

$$dI_x = y^2 dA = y^2\left(b - \frac{b}{h}y\right)dy = b\left(y^2 - \frac{y^3}{h}\right)dy$$

Fig. 3.9 A 2-D shape for
Example 3.3

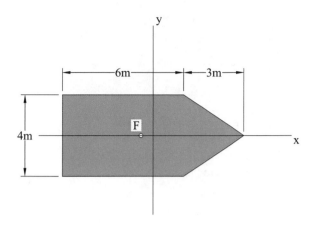

Thus the second moment of the area is

$$I_x = b \int_0^h \left(y^2 - \frac{y^3}{h} \right) dy = b \left[\frac{y^3}{3} - \frac{y^4}{4h} \right]_0^h = \frac{bh^3}{12}$$

We may also take the strips of width dx, parallel to the y-axis. Since dx is very small, the strip may be considered to be a rectangle. Its area is $y dx$ and its second moment about the x-axis is (from the second moment of a rectangle above)

$$dI_x = \frac{y^3 dx}{3} = \frac{1}{3} \left(-\frac{h}{b}x + h \right)^3 dx$$

When this is integrated along the x-axis from 0 to b, the same result as above is obtained.

Now, the second moment about the centroidal axis can be found by using the parallel axis theorem (remembering that the least second moment of an area is about its centroidal axis)

$$I_{xc} = I_x - \frac{1}{2}bh \left(\frac{h}{3} \right)^2 = \frac{bh^3}{12} - \frac{bh^3}{18} = \frac{bh^3}{36}$$

Example 3.3 Find the centroid of an area shown in Fig. 3.9 and then calculate the second moment of this area about the centroidal axis parallel to y-axis.

Solution
We can divide the area into a rectangle and a triangle. We can assume that the y-axis is at the mid-length of the shape, i.e. 4.5 m from the left end. Then, we see that the centroid of the rectangle is at $x = -1.5$ m and the centroid of the triangle is at $x = 2.5$ m. We can now construct a table as follows:

	Area (m^2)	Lever from y-axis (m)	First moment (m^2 − m)
Rectangle	24	−1.5	−36.0
Triangle	6	2.5	15.0
Total	**30**		**−21.0**

Therefore, the centroid F of the whole shape is at $x_c = \frac{-21}{30} = -0.7$ m. Here the negative sign indicates that the centroid is located at the left side of the y-axis.

The second moment of the rectangle about its own vertical centroidal axis is $\frac{4 \times 6^3}{12} = 18$ m^4 and the second moment of the triangle about its own vertical centroidal axis is $\frac{4 \times 3^3}{36} = 3$ m^4.

We can now apply the parallel axis theorem to find the second moment about the axis parallel to y-axis but passing through the point F. The clearest way of doing this is by creating another table as below.

	Second moment about own axis	Area (m^2)	Lever from F to own axis (m)	Area x lever^2 (m^4)	Second moment (m^4)
Rectangle	18.0	24.0	−0.8	15.36	33.36
Triangle	3.0	6.0	3.2	61.44	64.44
Total					**97.8**

We can see that the contribution of the triangle is largely due to its distance from F, while the contribution of the rectangle is due to its second moment about own axis and the separation.

We can arrive at the same result if we calculate the second moment of the rectangle and triangle about the y-axis first and then apply the parallel axis theorem to the whole shape to find the second moment about point F.

Example 3.4 A twin pontoon type semi-submersible has four circular columns of diameter 12 m. The transverse distance between the centres of the columns is 70 m. Calculate the transverse second moment of the waterplane area in its operational condition. Ignore all bracings. The second moment of area of a circle of diameter d about its diametrical axis is $\frac{\pi d^4}{64}$.

Solution

Semi-submersibles are designed to operate so that only its columns and associated structures (such as bracings) pierce the water surface. The area of the circle is $\pi d^2/4$. The distance from the centre of the circle to the longitudinal axis of the semi-submersible is $70/2 = 35$ m. The second moment of the circle about the longitudinal axis (centreline) can be found by using parallel axis theorem

$$I_T = \frac{\pi d^4}{64} + \frac{\pi d^2}{4}35^2 = 1017.88 + 138{,}544.24 = 139{,}562.12 \text{ m}^4$$

Since all four columns are of identical size and at the same distance away from the centreline,

$$I_T = 4 \times 139{,}562.12 = 558{,}248.48 \text{ m}^4$$

It can be seen that the contribution of the second moment about its own diametrical axis is very small (less than 1%) indeed. Nevertheless, the second moment is comparable to a monohull vessel despite its small waterplane area thanks entirely to the *separation*. As will be seen later, the transverse second moment of the waterplane is one of the main factors which influence the stability of a vessel. This explains why catamarans and semi-submersibles have very good initial stability characteristics.

3.4 Properties of 3-D Shapes

The first and second moments of area are usually about an axis. The first moment of volume and mass, on the other hand, is calculated about a plane. Their second moment, however, is calculated about an axis, as it is the inertia of the object in rotational motion about that axis (hence the alternative name, mass moment of inertia). Despite these little differences, the properties of volume and mass are really the same as those of 2-D shapes.

For instance, with an arbitrary object in space, the vertical mass moment about x-y plane is

$$M_{xy} = \int_M z \, dm$$

where

z is the vertical distance from the reference plane (x-y plane in this case) to the elemental mass dm, and M denotes the whole mass

Therefore the centroid of the object is at a point

$$\bar{z} = \frac{M_{xy}}{M} = \frac{\int_M z \, dm}{M}$$

above the reference plane. The same principle applies in the other directions. Note that volume can be treated in exactly the same way as mass. This is how the centres of gravity and buoyancy are calculated. For example, the centre of buoyancy can be found in the following manner.

Consider a Cartesian co-ordinate system where the origin is fixed at a point amidships on the centreline on the water plane. Positive x is forward, y is positive starboard and z is positive upward, maintaining the right-hand rule. Then the moments of the underwater volume about the x-y and y-z planes are

$$M_{xy} = \int_V z dv \text{ and } M_{yz} = \int_V x dv$$

where

V is the whole underwater volume and dv is the elemental volume

The longitudinal position of the centre of buoyancy (LCB) is M_{yz}/V from the midship and its vertical position (VCB) is M_{yz}/V from the waterline, remembering the sign convention.

The mass moment of inertia of an object is calculated about an axis, as mentioned before. For example, the moment of inertia about x-axis is

$$I_{xx} = \int_M \left(y^2 + z^2\right) dm$$

Note that $\left(y^2 + z^2\right)$ is the squared distance of dm from the x-axis.

Movement of Centroid Due to Addition or Subtraction of Mass or Volume

Consider an object, with mass m_1 and the centroid l_1 from a reference plane. Some mass m_2 is added to it so that its centroid is at l_2 from the plane. Then the distance of the new centroid from the reference plane, l, is

$$l = \frac{m_1 l_1 + m_2 l_2}{m_1 + m_2}$$

This is quite obvious from the principles that

- the distance to the centroid can be obtained by dividing the first moment by the mass; and
- the moment of an object about an axis can be found by summing the moments of its component parts about the axis.

Of course, when m_2 is subtracted from m_1, then the sign of m_2 will be negative.

Sometimes it is more convenient to know how much the centre of gravity moves in this situation. The movement of the centroid (from l_1, that is), dl is

$$dl = l - l_1 = \frac{m_1 l_1 + m_2 l_2}{m_1 + m_2} - l_1 = \frac{m_1 l_1 + m_2 l_2 - m_1 l_1 - m_2 l_1}{m_1 + m_2}$$
$$= \frac{m_2 l_2 - m_2 l_1}{m_1 + m_2} = \frac{m_2 (l_2 - l_1)}{m_1 + m_2}$$

The numerator of the last is the mass moment of m_2 about the plane passing through the centre of gravity of m_1 and parallel to the original reference plane.

This idea can be applied to the situation where parts of a given object are moved to new locations. This case can be regarded as some parts removed and the same parts added elsewhere. Some worked examples will illustrate this very much more clearly.

Example 3.5 A ship of 10,000 tonnes mass has its centre of gravity at 7 m above baseline. Some cargo of mass 1000 tonnes is loaded onto the ship so that the added cargo's centre of gravity is at 10 m above baseline. Calculate the new vertical position of the ship's centre of gravity (VCG).

Solution

$$\text{The new } KG = \frac{10000 \times 7 + 1000 \times 10}{10000 + 1000} = 7.273 \, \text{m}$$

We may get the same result by taking the moment about the original centre of gravity, thus

$$\text{the new CG relative to the original CG} = \frac{10000 \times 0 + 1000 \times (10 - 7)}{10000 + 1000} = 0.273 \, \text{m}$$

Example 3.6 A ship of 8000 tonnes mass has its centre of gravity at 4 m aft of midship. Some cargo of mass 500 tonnes is moved forward by 20 m. Calculate the longitudinal position of the centre of gravity (LCG).

Solution
We may regard the movement as a combination of two operations: unload 500 tonnes from a location x, say, and load 500 tonnes at a location x + 20 m. Taking fwd of midship as positive,

$$\text{The new LCG} = \frac{8000 \times (-4) - 500 \times x + 500 \times (x + 20)}{8000 - 500 + 500}$$
$$= \frac{8000 \times (-4) + 500 \times 20}{8000} = -2.75m$$

i.e., 2.75 m aft of midship. The important thing to note here is that when a mass is simply moved, we do not need to do this in two operations. All we have to do is add the moment of the movement. Note, however, that the total mass has not changed.

Example 3.7 A ship of mass displacement 10,000 tonnes has its centre of gravity at 8 m above baseline. The following operations are carried out (AB means above base):

Load   500 tonnes   3 m AB
Discharge 300 tonnes   5 m AB
Load   200 tonnes   12 m AB
Move   1000 tonnes downward by 2 m

Calculate new VCG.

Solution

$$\text{The new VCG} = \frac{10000 \times 8 + 500 \times 3 - 300 \times 5 + 200 \times 12 - 1000 \times 2}{10000 + 500 - 300 + 200} = 7.731 \, \text{m}$$

In fact when a number of operations are carried out like this, it is best to compile a table, something like this

Operation	Mass (tonnes)	v.c.g. (m)	dΔ	Moment
Original ship	10,000	8	10,000	80,000
Load	500	3	+500	1500
Discharge	300	5	−300	−1500
Load	200	12	+200	2400
Move	1000	by −2	0	−2000
		Σ	**10,400**	**80,400**

$$\text{New VCG} = \frac{80400}{10400} = 7.731 \, \text{m}$$

Key Points

- The position of the centroid of a volume or mass can be obtained by dividing its first moment about a given axis by its volume or mass.
- The movement the centroid due to addition to or subtraction from a volume or mass can be found by taking the first moment about the original centroid and apply the above principle.

Excercises

1. Find the first moments of area about x and y axes for the various shapes shown below, and determine the location of their centroid.

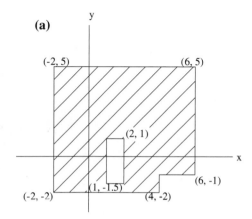

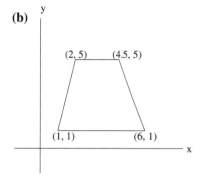

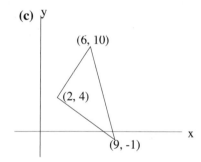

2. For the waterplane shown below calculate

first moment of area about midship and hence longitudinal position of its centroid from midship;
second moments of area about longitudinal and transverse axes through the centroid.
The second moment of area of a complete circle about its diametrical axis is $\frac{1}{4}\pi r^4$ and the centroid of a semi-circle is at $\frac{4r}{3\pi}$ away from the diametrical axis.

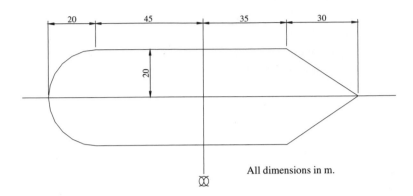

20   45   35   30

20

All dimensions in m.

3. Calculate the second moment of area of a catamaran shape below about the longitudinal and transverse axes through its centroid.

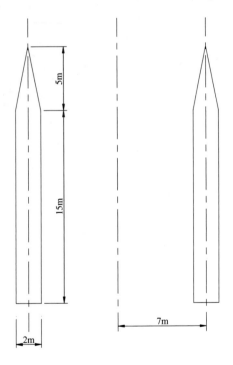

5m

15m

2m

7m

Chapter 4
Numerical Integration Methods

4.1 Introduction

Not surprisingly, integration in various forms keeps appearing in hydrostatics, since area is calculated by integrating line lengths and volume by integrating sectional areas and so on. It is hardly an exaggeration to say that most of hydrostatic calculations were based on moment theories and integration of one sort or another.

When we are dealing with geometrical shapes which can be readily expressed in a mathematical form (particularly polynomials), the integration in most cases can be accomplished by manipulating the mathematical expression. However, the parametric piecewise polynomials often used to express hull forms are in general more difficult to integrate. In any case, naval architects of bygone era before computers arrived had to devise some methods to do the integration manually. A number of numerical integration techniques which came into general use in ship design offices are used extensively even now. These numerical integration methods produce approximate answers, although sufficiently accurate for some engineering purposes. Indeed ingenious, albeit somewhat clumsy to use requiring infinite care and patience, devices called 'integrators' were invented to carry out 'accurate' integration, but thankfully these have disappeared from ship design offices, as computers can produce more reliable results in a much shorter time.

In this part, we will explore some of the more popular numerical integration techniques. Before doing so, however, it will be useful to summarise various ways of finding some hydrostatic quantities as shown in Table 4.1.

© Springer Nature Singapore Pte Ltd. 2019
B. S. Lee, *Hydrostatics and Stability of Marine Vehicles*, Springer Series on Naval Architecture, Marine Engineering, Shipbuilding and Shipping 7,
https://doi.org/10.1007/978-981-13-2682-0_4

Table 4.1 Some hydrostatic quantities and how they are obtained

To get	Integrand	Direction
WP area	Breadths	Longitudinally
Long'l moment of WP	Long'l moments of breadth	Longitudinally
Section area	Breadths	Vertically
Volume displacement	WP area	Vertically
	Section area (SA)	Longitudinally
Vertical moment of displacement	Vertical moment of SA	Longitudinally
	Vertical moment of WPA	Vertically
Long'l moment of displacement	Long'l moment of SA	Longitudinally
	Long'l moment of WPA	Vertically
I_T of WP	Transverse second moment of breadths	Longitudinally
I_L of WP	Longitudinal second moment of breadths (about midship)	Longitudinally (and then apply parallel axis theorem)

4.2 Trapezoidal Rule

Consider a curve A–D as shown in Fig. 4.1a with the known points of (x_i, y_i), $i = 0$, 1, 2, 3, ... n, and we would like to calculate the area ABCD. Note that point (x_0, y_0) is at A and (x_n, y_n) is at D. The curve A-D can be assumed to be made of n straight line segments. Then an approximate area of ABCD can be found by simply summing the areas of the trapezoids thus formed as shown in Fig. 4.1b.

Thus area ABCD =

$$\frac{1}{2}\left[(x_1 - x_0)(y_0 + y_1) + (x_2 - x_1)(y_1 + y_2) + \cdots + (x_n - x_{n-1})(y_{n-1} + y_n)\right]$$

If $x_1 - x_0 = x_2 - x_1 = \cdots = h$, i.e. x_i are evenly spaced on the horizontal axis, the above can be reduced to

$$\frac{1}{2}h\left[(y_0 + y_1) + (y_1 + y_2) + \cdots + (y_{n-1} + y_n)\right] = \frac{h}{2}(y_0 + 2y_1 + 2y_2 + \cdots + y_n)$$

The accuracy of this technique depends largely on how closely the curve can be represented as a collection of straight lines. It is obvious, therefore, that the more points on the curve are known, the more accurate the result will be. The method is extremely simple to understand and use, and despite its somewhat crude approximation can often produce sufficiently accurate results for some applications. Indeed some very sophisticated computer programs sometimes rely on this method.

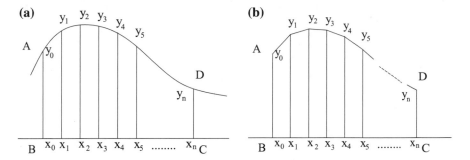

Fig. 4.1 **a** Area ABCD to be calculated. **b** Simplified to trapezoids

Trapezoidal rule is one of the Newton-Cotes formulae for numerical integration which cover polynomials of any order, trapezoidal being for linear polynomial approximation. For any curve other than straight line, approximating it with a higher order curve will produce more accurate area, and for this we turn to Simpson's rules which are again special cases of Newton-Cotes formulae.

4.3 Simpson's Rules

Simpson's rules are the most widely used numerical integration techniques because of their simplicity and wide applicability. Consider a curve A-D shown in Fig. 4.2 with three known points the x values of which are such that $x_2 - x_1 = x_3 - x_2$ (in other words evenly spaced). For ease of calculation we will put the y-axis at $x = x_2$, thus making $x_2 = 0$. Let $x_1 = -h$, then $x_3 = h$.

We now assume that the curve A–D can be expressed as a cubic polynomial of x, thus

$$y = a_0 + a_1 x + a_2 x^2 + a_3 x^3 \qquad (4.1)$$

The area ABCD, say A, can be found as

$$A = \int_{-h}^{h} y\,dx = \left[a_0 x + \frac{1}{2}a_1 x^2 + \frac{1}{3}a_2 x^3 + \frac{1}{4}a_3 x^4 \right]_{-h}^{h} = 2a_0 h + \frac{2}{3}a_2 h^3 \qquad (4.2)$$

Since

$$y = y_2 \text{ when } x = 0,\ a_0 = y_2 \qquad (4.3)$$

$$y = y_1 \text{ when } x = -h,\ y_1 = a_0 - a_1 h + a_2 h^2 - a_3 h^3 \qquad (4.4)$$

$$y = y_3 \text{ when } x = h,\ y_3 = a_0 + a_1 h + a_2 h^2 + a_3 h^3 \qquad (4.5)$$

Fig. 4.2 Simpson's First
Rule

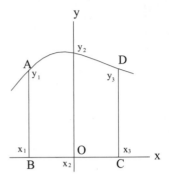

Adding (4.4) to (4.5) and solving the result for a_2 gives

$$a_2 = \frac{\frac{1}{2}(y_1 + y_3) - y_2}{h^2}$$

Substituting this and (4.3) into (4.2) gives

$$\text{area ABCD} = 2y_2 h + \frac{2}{3}h^3 \left(\frac{\frac{1}{2}(y_1 + y_3) - y_2}{h^2} \right) = \frac{h}{3}(y_1 + 4y_2 + y_3)$$

This is known as the **Simpson's First Rule** and a moment's reflection will show that as long as three points on the curve, the x-values of which are evenly spaced, are known, this rule can be applied regardless of what those x-values are (remember we derived the above for $x_2 = 0$, but x-values do not appear in the rule except the 'common interval' h). This rule is sometimes known as Simpson's 1/3 rule or 1-4-1 rule for obvious reasons.

Note that this rule produces a precise value for any quadratic or cubic polynomials. Therefore, the accuracy of this method depends largely on how closely the curve can be approximated as a quadratic or cubic polynomial. If, on the other hand, the curve contains segments which can only be expressed in polynomials of different orders, for example a cubic curve adjoining a straight line segment, the accuracy will decrease.

If we have to deal with 5 points on the curve with the x-values equally distributed as shown in Fig. 4.3, then we can divide the area into two segments ABCE and ECDF, and apply Simpson's first rule to each portion before summing them, thus

$$\text{area ABFD} = \frac{h}{3}(y_1 + 4y_2 + y_3) + \frac{h}{3}(y_3 + 4y_4 + y_5) = \frac{h}{3}(y_1 + 4y_2 + 2y_3 + 4y_4 + y_5)$$

The numbers (1, 4, 2, 4, 1) are known as Simpson multipliers.

If there is a half station at the beginning—for example, 0, ½, 1, 2, 3, 4, 5, the integration can be performed by applying the rule separately to the segments 0, ½, 1

Fig. 4.3 Five equally spaced ordinates

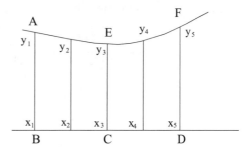

and 1, 2, 3, 4, 5. Of course, the common interval for the first segment will be half of that for the latter. Thus,

$$\text{area} = \int_{x_1}^{x_3} y\,dx = \frac{h/2}{3}(y_0 + 4y_{1/2} + y_1) + \frac{h}{3}(y_1 + 4y_2 + 2y_3 + 4y_4 + y_5)$$

$$= \frac{h}{3}(0.5y_0 + 2y_{1/2} + 0.5y_1 + y_1 + 4y_2 + 2y_3 + 4y_4 + y_5)$$

$$= \frac{h}{3}(0.5y_0 + 2y_{1/2} + 1.5y_1 + 4y_2 + 2y_3 + 4y_4 + y_5)$$

Therefore, the Simpson multipliers in this case will be (0.5, 2, 1.5, 4, 2, 4, 1). One often encounters numerical integration in a tabular form and a sample of this is given in Examples 4.1 and 4.2.

Simpson's First Rule, as we have seen, applies to three equally spaced ordinates. There is another Simpson's rule which applies to four equally spaced ordinates, and it is known as **Simpson's Second Rule** (sometimes known as Simpson's 3/8 rule or 1-3-3-1 rule). With the four ordinates y_1, y_2, y_3 and y_4 at x_1, x_2, x_3 and x_4, and the common interval h,

$$\text{area} = \int_{x_1}^{x_4} y\,dx = \frac{3h}{8}(y_1 + 3y_2 + 3y_3 + y_4)$$

You can work out how this may be so and prove that it produces accurate results for cubic polynomials quite easily.

Example 4.1 Calculate the area and the longitudinal position of the centroid of a waterplane, half breadths of which are shown as ordinates in Table 4.2. The length of this waterplane is 150 m.

Solution
It can be seen from Table 4.1 that the integrand for the longitudinal moment of the waterplane area is the longitudinal moment of the breadths, i.e. breadths multiplied

Table 4.2 Solution of Example 4.1

	①	②	③ = ① × ②	④	⑤ = ③ × ④
Station	Half-breadths	SM	$f(A)$	Lever	$f(M_L)$
0	10.00	0.25	2.5	−5.0	−12.5
¼	10.61	1	10.61	−4.75	−50.3975
½	11.30	0.5	5.65	−4.5	−25.425
¾	12.07	1	12.07	−4.25	−51.2975
1	12.94	0.75	9.705	−4.0	−38.82
1½	14.92	2	29.84	−3.5	−104.44
2	17.25	1.5	25.875	−3.0	−77.625
3	22.94	4	91.76	−2.0	−183.52
4	30.00	2	60.0	−1.0	−60.0
5	30.00	4	120.0	0	0.0
6	30.00	2	60.0	1.0	60.0
7	19.13	4	76.52	2.0	153.04
8	10.50	1.5	15.75	3.0	47.25
8½	7.03	2	14.06	3.5	49.21
9	4.13	0.75	3.0975	4.0	12.39
9¼	2.88	1	2.88	4.25	12.24
9½	1.78	0.5	0.89	4.5	4.005
9¾	0.82	1	0.82	4.75	3.895
10	0.00	0.25	0	5.0	0
		$\Sigma f(A) =$	**542.0275**	$\Sigma f(M) =$	**−261.995**

by the x-coordinate of each station with the transverse axis at midship. There are 11 whole stations, and therefore the common interval between two consecutive whole stations is $h = 150/10 = 15$ m. The lever, or x-coordinates, of the stations can be expressed as $l = (stn - 5) \times h$. So, we can simply use $(stn - 5)$ as the lever while remembering to multiply the end total with h. The Simpson's formula is then applied as shown in Table 4.2.

$$\text{Thus, the area} = 2 \times \frac{h}{3} \times f(A) = 2 \times \frac{15}{3} \times 542.0275 = 5420.275 \text{ m}^2$$

$$\text{moment about midship} = 2 \times \frac{h}{3} \times h \times \sum f(M)$$

$$= 2 \times \frac{15^2}{3} \times (-261.995) \times 15 = 39,299.25 \text{ m}^3$$

longitudinal position of the centroid is

$$x_c = 2 \times \tfrac{h}{3} \times h \times \sum f(M) \bigg/ 2 \times \tfrac{h}{3} \times \sum f(A) = \frac{h \times f(M)}{f(A)} = 7.250 \text{ m}$$

It is crucial to remember that the ordinates represent the values of the integrands shown in Table 4.1.

Example 4.2 Calculate the area and the longitudinal position of the centroid of a waterplane 60 m long, the half breadths of which are shown in the table below. Then calculate the longitudinal second moment of this area about the centroid and the transverse second moment about the centreline.

Station	Half breadths (m)
0	3.41
1	5.20
2	6.25
3	6.60
4	6.32
5	5.05
6	1.71

Solution
The area and the centroid can be found in the same way as in Exercise 4.1. The two second moments need a little explaining first using Fig. 4.4.

Fig. 4.4 A strip of infinitesimally small thickness

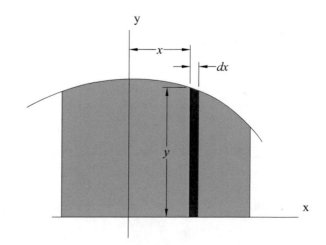

Longitudinal Second Moment
We first consider the second moment about the transverse axis at midship first. The second moment of the infinitesimally thin strip of thickness dx about y-axis at midship is

$$dI_m = ydx \cdot x^2.$$

The second moment of the whole area, therefore, is

$$I_m = 2 \int_{x_0}^{x_6} yx^2 dx$$

It is clear from this that the integrand is yx^2. Of course, in applying the Simpson's Rule, we can use x/h as the lever instead of x, always remembering to multiply the result with h^2. Having obtained I_m, we can use the parallel axis theorem to find the second moment about the centroidal axis.

Transverse Second Moment
The thin strip can be regarded as a rectangle. As discussed in Sect. 3.3, the second moment of this strip about the x-axis is

$$dI_T = y^3 dx /3$$

The transverse second moment of the whole area then is

$$I_T = \frac{2}{3} \int_{x_0}^{x_6} y^3 dx$$

The integrand in this case is y^3.
We are now ready to compile a table for Simpson's Rule as follows:

Station	y	SM	Lever	Lever^2	f(A)	f(M_m)	f(I_m)	f(I_T)
0	3.41	1	−3	9	3.41	−10.23	30.69	39.65
1	5.2	4	−2	4	20.8	−41.6	83.2	562.43
2	6.25	2	−1	1	12.5	−12.5	12.5	488.28
3	6.6	4	0	0	26.4	0	0	1149.98
4	6.32	2	1	1	12.64	12.64	12.64	504.87
5	5.05	4	2	4	20.2	40.4	80.8	515.15
6	1.71	1	3	9	1.71	5.13	15.39	5.00
				Total	97.66	−6.16	235.22	3265.37

From this we can get the following:

$$A = \frac{2h}{3} \sum f(A) = 651.067 \text{ m}^2$$

$$M_{ms} = \frac{2h}{3} h \sum f(M) = -410.667 \text{ m}^3$$

Longitudinal position of centroid (LCF) $= M_{ms} / A = -0.631$ m

$$I_L = \frac{2h}{3} h^2 \sum f(I_{ms}) - A \times LCF^2 = 156{,}554.3 \text{ m}^4$$

$$I_T = \frac{2h}{3} \frac{h^2}{3} \sum f(I_{ms}) = 7256.382 \text{ m}^4$$

4.4 Other Numerical Integration Methods

Although Simpson's rules are enough in most cases encountered in naval architecture, there are a few other numerical integration methods which can be useful in certain situations. We shall have a look at Tchebycheff's rules here.

Tchebycheff's Two-Ordinate Rule

As with the Simpson's rules, we start with a function which is expressed as a cubic polynomial of x as in Eq. (4.1) for which the definite integral from $x = -h$ to $x = h$ is given by Eq. (4.2).

We now assume that there is a value $x = x_1$ and $x_2 = -x_1$ such that Area ABCD can be expressed as $M(y_1 + y_2)$,

where

$$y_1 = f(x_1) = a_0 + a_1 x_1 + a_2 x_1^2 + a_3 x_1^3$$

and

$$y_2 = f(x_2) = a_0 - a_1 x_1 + a_2 x_1^2 - a_3 x_1^3$$

Substituting these into the expression of area,

$$\text{Area} = M(2a_0 + 2a_2 x_1^2).$$

Comparing this with Eq. (4.2) gives

$$M = h$$

$$Mx_1^2 = \frac{1}{3} h^3$$

Therefore,

$$x_1 = \pm \frac{h}{\sqrt{3}}$$

In other words the area under the curve from $x = -h$ to h can be obtained by multiplying h with the sum of two ordinates at $x = -0.57735\,h$ and $x = 0.57735\,h$. This is for a special case where the base is divided equally about the y-axis. In general, the first ordinate can be taken at x_1 which is $0.42265\,h$ from the left boundary of the base and the second ordinate similarly from the other end.

Tchebycheff's Three-Ordinate Rule

Again we start with Eqs. (4.1) and (4.2), but this time we set up three ordinates at $x = x_1$, 0 and $-x_1$, with their respective values of y_1, y_2 and y_3. Again we express the area ABCD as $M(y_1 + y_2 + y_3)$, and we know

$$y_1 = f(x_1) = a_0 + a_1 x_1 + a_2 x_1^2 + a_3 x_1^3$$

$$y_2 = f(0) = a_0$$

$$y_3 = f(-x_1) = a_0 - a_1 x_1 + a_2 x_1^2 - a_3 x_1^3$$

Substituting these into the expression for the area and comparing it with Eq. (4.2) gives

$$\text{Area} = M\left(3a_0 + 2a_2 x_1^2\right) = 2a_0 h + \frac{2}{3} a_2 h^3$$

Therefore $M = \frac{2}{3}h$ and $x_1 = \pm \frac{1}{\sqrt{2}}h$

In other words the first and the third ordinates are at $x = \pm 0.707107\,h$. The area is then $\frac{2h}{3}(y_1 + y_2 + y_3)$.

We have seen two of Tchebycheff's rules, but in fact there is a series of them for various numbers of ordinates and corresponding order of polynomials. For us, however, what we have examined here suffices.

4.5 Polar Integration

In some cases of naval architecture we have to deal with areas enclosed by polar curves as shown in Fig. 4.5. The numerical integration techniques discussed earlier in this chapter can also be used for polar integration, but it will be useful to discuss how polar integration is carried out first.

Fig. 4.5 Polar integration

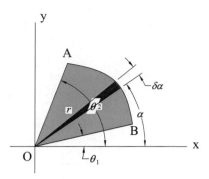

Consider an area represented by two lines OA and OB and a curve segment AB as shown in Fig. 4.5. The angle of the line OA to the x-axis is θ_2 and OB is θ_1 away from the x-axis. We construct an elemental area dA of angle $d\alpha$ as shown at an angle α from the x-axis. Since this area can be regarded as a triangle with the height of $rd\alpha$ and base of r,

$$dA = \frac{1}{2}rrd\alpha = \frac{1}{2}r^2 d\alpha$$

Therefore, the area is

$$A = \int_{\theta_1}^{\theta_2} \frac{1}{2}r^2 d\alpha$$

Also, the first moments of the elemental area about the x-axis and y-axis are

$$dM_x = \frac{1}{2}r^2 d\alpha \cdot \frac{2}{3}r \sin\alpha = \frac{1}{3}r^3 \sin\alpha d\alpha$$

and

$$dM_y = \frac{1}{2}r^2 d\alpha \cdot \frac{2}{3}r \cos\alpha = \frac{1}{3}r^3 \cos\alpha d\alpha$$

Therefore,

$$M_x = \int_{\theta_1}^{\theta_2} \frac{1}{3}r^3 \sin\alpha d\alpha$$

$$M_y = \int_{\theta_1}^{\theta_2} \frac{1}{3} r^3 \cos \alpha \, d\alpha$$

The integrands can be easily identified from these. You can work out how these integrals can be executed with the numerical integration methods discussed earlier in this chapter.

This type of integration can be useful in the calculation of rotational stability, as will be seen in the study of transverse stability.

Key Points

- Simpson's integration rules can be applied accurately to any curve which can be approximated as a cubic polynomial.
- Simpson's first rule can be applied to $(3+2n)$ evenly spaced ordinates where $n = 0, 1, 2 \dots$.
- Simpson multipliers (first rule) for 7 evenly spaced ordinates, for example, are $(1, 4, 2, 4, 2, 4, 1)$.
- The area under a curve can be found by using the first rule with $\frac{h}{3} \sum f(A)$.
- Simpson's second rule can be applied to $(4+3n)$ evenly spaced ordinates where $n = 0, 1, 2 \dots$.
- Simpson multipliers (second rule) for 10 evenly spaced ordinates, for example, are $(1, 3, 3, 2, 3, 3, 2, 3, 3, 1)$.
- The area under a curve can be found by using the second rule with $\frac{3h}{8} \sum f(A)$.

Excercises

1. A curve has the following ordinates spaced 1.68 m apart: 10.86, 13.53, 14.58, 15.05, 15.24, 15.28, 15.22. Calculate the area under this curve using

    (a) Simpson's first rule
    (b) Simpson's second rule
    (c) the trapezoidal rule.

2. A curvilinear figure has the following ordinates at equidistant intervals: 12.4, 27.6, 43.8, 52.8, 44.7, 29.4 and 14.7.
    Calculate the percentage difference from the area found by Simpson's first rule when finding the area by (a) Simpson's second rule and (b) the trapezoidal rule. Explain this discrepancy.

3. The half-ordinates of a vessel 144 ft between perpendiculars are given below:
    In addition there is an appendage 21.6 ft long abaft the AP whose half-ordinates at equally spaced intervals are respectively, 0.0, 9.6, 14.0, 17.0 (AP).
    Find the area and position of the centre of the area of the complete waterplane.

Ord No	AP(0)	1	2	3	4	5	6	7	FP(8)
½ ord (ft)	17.0	20.8	22.4	22.6	21.6	18.6	12.8	5.6	0

4. The cross sectional areas of a ship, 72 m LBP, 11.5 m beam and 4.3 m draught, are as follows:

Station	0	½	1	1½	2	3	4	5	6
CSA (m²)	0	17.1	28	35	40	45	46	46	46
Station	7	8	8½	9	9¼	9½	9¾	10	
CSA (m²)	45	31	24	17	14.2	8	4.9	0	

Determine the volume of displacement, mass of displacement in salt water, LCB, C_B, C_P, and C_M.

5. The cross sectional areas of a ship, 72 m LBP, 11.5 m beam and 4.3 m draught, are as follows:

Station	0	1	2	3	4	5	6	7	8	9	10
CSA (m²)	15	30	40	45	46	46	46	45	31	17	0

The waterplane at 4.3 m draught has half-breadths as follows:

Station	0	1	2	3	4	5	6	7	8	9	10
Half-breadth (m)	2.5	3.9	5.0	5.75	5.75	5.75	5.75	5.75	5.0	3.2	0.0

Calcualte the following items:
volume and mass displacement (sea water)
C_B, C_P, C_W and C_M
LCB (longitudinal centre of buoyancy)
LCF (longitudinal centroid of the waterplane)
I_L (longitudinal second moment of WP area about axis through LCF)
I_T (transverse second moment of WP area about the centreline).

Chapter 5
Trim and Longitudinal Stability

Stability of a ship is defined as its ability to return to the normal operating attitude when disturbed from it by transitory forces or moments. This concept will be explored in more detail in the next chapter, but here it is worthwhile to note that most ships have sufficient longitudinal stability in their intact condition and, therefore, it can be considered to be of little interest to us in terms of safety. However, the longitudinal stability affects the trim of the vessel directly, and trim is a very important factor in determining the operational efficiency of ships. We shall, therefore, briefly examine here the longitudinal stability with emphasis on trim. There are three conditions of trim:

even or level keel
trim by the stern
trim by the head or bow

The formal definition of trim is the difference between the draught at FP and draught at AP, i.e.

$$t = T_f - T_a$$

where t is trim, T_f is the draught at FP and T_a is the draught at AP.

It can be seen therefore that a positive trim indicates trim by the bow and a negative trim means trim by the stern.

© Springer Nature Singapore Pte Ltd. 2019
B. S. Lee, *Hydrostatics and Stability of Marine Vehicles*, Springer Series on Naval Architecture, Marine Engineering, Shipbuilding and Shipping 7,
https://doi.org/10.1007/978-981-13-2682-0_5

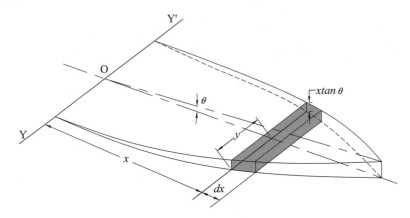

Fig. 5.1 Forward submerged wedge when the ship trims

5.1   Longitudinal Centre of Floatation (LCF)

Consider Fig. 5.1 which represents the forward wedge which submerges when the vessel changes trim (bow down) without altering its displacement.

The volume of an elemental strip of half breadth y and length δx is $2yx \tan\theta\, \delta x$. Therefore, the total volume of the wedge is

$$2\tan\theta \int_{0}^{L_f} xy dx.$$

But $2\int_{0}^{L_f} xy dx$ is the first moment of the original waterplane about YOY'.

Similarly for the aft portion of the vessel, the volume is $-2\tan\theta \int_{-L_a}^{0} xy dx$.

(Note that L_a and L_f are the length aft and forward respectively from YOY'.)

Since the vessel has not changed its displacement, the volumes of these two wedges should be the same, i.e.

$$2\tan\theta \int_{0}^{L_f} xy dx = -2\tan\theta \int_{-L_a}^{0} xy dx$$

Note that the integrals on oth sides are half of the first moment of the fore and aft waterplanes respectively. Therefore,

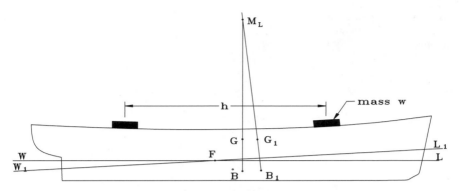

Fig. 5.2 Change in trim due to movement of cargo

$$2 \tan \theta \int_{-L_a}^{L_f} xy\,dx = 0$$

In other words, the first moments of the original waterplane (about the axis YOY')
fore and aft are the same in magnitude although opposite in sign. This shows that
in order for the vessel to change trim whilst maintaining the original displacement
volume, the trim axis YOY' should be such that the first moment of the original
waterplane about this axis is zero. Put in another way, YOY' must pass through the
centroid of the waterplane, and this point is called the **centre of flotation (CF)**. The
longitudinal position of CF is often represented by the distance of this point from the
midship and denoted as LCF. Negative LCF means CF is located aft of the midship.

5.2 Moment to Change Trim (MCT)

Consider Fig. 5.2 which shows a vessel floating at waterline WL. It is in equilibrium
because the centre of gravity (G) is vertically above the centre of buoyancy (B). A
piece of cargo of mass w tonnes is then moved longitudinally through a distance h as
shown. When the cargo is moved, the mass of the ship does not change but the CG of
the vessel moves from G to G_1. This will cause the vessel to change trim about LCF
(point F) and finally settle at waterline W_1L_1 in such a way that, after trimming, the
CB will move to a point B_1 vertically below G_1 (since that is the new equilibrium).

Before we start investigating the mechanics of this situation, we must declare
that, in most cases where there is no possibility of confusion, 'moment' means 'mass
moment' rather than 'force moment'.

If the ship is trimmed momentarily like this by a transitory moment and not by the
movement of a cargo, the moment causing the trim must have been the same as the
'righting' moment of the vessel at W_1L_1. In this case, the centre of gravity is still at G,

Fig. 5.3 Gravitational and
buoyancy forces for a small
trimming angle

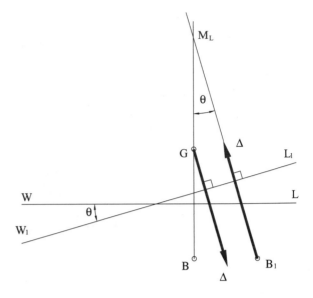

while the centre of buoyancy has moved to B_1. The moment that the ship experiences
in this condition is the displacement multiplied by the lever between buoyancy and
gravitational force. It can be seen from Fig. 5.2 that the lever is $GM_L \tan\theta$, and
therefore the righting moment, i.e. the moment which tries to put the ship in the trim
before the disturbance occurred, is $\Delta GM_L \tan\theta$. If the **change in trim** is t and the
length of the vessel is L, then $\tan\theta = t/L$. We can see that the moment to alter trim
by t is therefore $\Delta GM_L \, t/L$. From this we can obtain the moment to change trim by
1 cm or **MCT** as

$$MCT = \frac{\Delta GM_L}{100\,L}\text{ tonne} - \text{m/cm}.$$

5.3   Longitudinal Metacentric Height

The formula for MCT derived above is very useful in estimating the attitude of a ship
after loading and unloading operation. In order to complete the formula, however,
we need to calculate the longitudinal metacentric height (GM_L).

If a ship floating at WL changes its attitude so that the new waterline is W_1L_1 due
to a temporary external moment as shown in Fig. 5.2, the centre of buoyancy moves
to B_1, while the centre of gravity stays at the original position G. The force diagram
in this situation is shown in Fig. 5.3. For a very small trim the point of intersection
between the vertical (to WL) line from B and the vertical (to W_1L_1) line from B_1 is
known as the longitudinal metacentre (M_L).

It can be seen that $BB_1 = BM_L \tan\theta$.

The centre of buoyancy moves to B_1, because some portion of the underwater volume has been transferred from aft to forward. Referring to the discussion in Sect. 5.1, the longitudinal moment of volume transfer is

$$\text{moment} = \int_{-L_a}^{L_f} \left(\text{elementary volme}\right) \times x$$

$$= \int_{-L_a}^{L_f} (2yx \tan\theta\, dx) \times x$$

$$= 2\tan\theta \int_{-L_a}^{L_f} yx^2\, dx$$

The last item in the above expression is actually the longitudinal second moment of the waterplane, I_L, multiplied by $\tan\theta$. Also the shift in centre of buoyancy, BB_1, can be obtained by dividing the moment by the total volume.

Therefore,

$$BB_1 = I_L \tan\theta / \nabla = BM_L \tan\theta$$
$$\therefore BM_L = I_L / \nabla$$

From this

$$GM_L = KB + BM_L - KG$$

However, some people prefer to use BM_L instead of GM_L, since BM_L is nearly 100 m even for fairly small ships, while $(KB - KG)$ tends to be a few metres. Nevertheless, wherever possible, GM_L should be used.

5.4   Changes in Loading Condition

Consider a vessel initially floating in equilibrium. Several masses $(m_1, m_2, ..., m_n)$ are loaded/discharged with their centroid at distances $(x_1, x_2, ..., x_n)$ from the original LCG. We wish to find the final draught and trim when loading/discharging is complete.

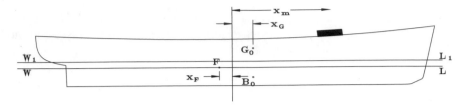

Fig. 5.4 Locations of key points

A table can be prepared something like the following:

Masses	Lever (from LCG)	Moment
$+m_1$	$+x_1$	$+m_1x_1$
$-m2$	$+x_2$	$-m_2x_2$
$-m3$	$-x_3$	$+m_3x_3$
$+m4$	$-x_4$	$-m_4x_4$
...	...	...
...	...	...
–		–
$\Sigma\, m$		$\Sigma\, mx$

Note + for masses added, fwd of LCG, trim by the bow
– for masses discharged, aft of LCG, trim by the stern

This sign convention is by no means universal, and indeed some people do not use the signs but rely on 'fwd' and 'aft', but it is much neater and less prone to mistakes if the signs are used, as long as the convention is strictly adhered to.

If the original displacement was Δ_0,

$\Delta = \Delta_0 + \Sigma m$ and

new LCG $= (LCG)_0 + \Sigma mx/\Delta$

LCB can be obtained from the table of hydrostatics at the displacement Δ. This will also give the even keel draught and BM_L.

Trimming moment $= \Delta \times$ (distance between LCB and LCG) and

Trim $=$ (trimming moment)$/(100 \times MCT)$ (m).

For conventional merchant vessels, there is a slightly simpler way of calculating the trimming moment without having to find the new centres of gravity and buoyancy.

Consider a vessel originally floating at the waterline WL as shown in Fig. 5.4. In this condition the centres of gravity and buoyancy of the vessel (G_0 and B_0) are vertically in line, since the vessel is in equilibrium. Let x_G be the distance of G_0 from the midship. Note that the distance from B_0 to midship will also be x_G. A piece of cargo of mass m is loaded at a distance x_m from midship as shown in the diagram.

We shall consider what happens to the ship in two stages: parallel sinkage due to the added mass displacement; and trim due to the longitudinal shift of the centres of gravity and buoyancy.

If the ship is not allowed to change the trim, the added mass m causes the ship to increase its draught to W_1L_1. The mass of water displaced by the newly immersed layer between WL and W_1L_1 should be the same as m, and its longitudinal centroid can be said to be approximately at the LCF of WL for relatively small m. If the LCF is at distance x_F from the midship, then the centre of buoyancy moves to B_1 which is at distance x_B from the midship, where

$$x_B = \frac{\Delta \cdot x_G + m \cdot x_F}{\Delta + m}$$

Let the longitudinal distance of the new centre of gravity (G_1) be x'_G from the midship. Then

$$x'_G = \frac{\Delta \cdot x_G + m \cdot x_m}{\Delta + m}$$

As we have seen above, the trimming moment $= \Delta *$ (distance between LCB and LCG), where the distance between LCB and LCG is

$$x'_G - x_B = \frac{\Delta \cdot x_G + m \cdot x_m - (\Delta \cdot x_G + m \cdot x_F)}{\Delta + m} = \frac{m(x_m - x_F)}{\Delta + m}$$

The trimming moment, therefore, is

$$(\Delta + m) \frac{m(x_m - x_F)}{\Delta + m} = m(x_m - x_F)$$

This method allows the trimming moment to be calculated without having to find new LCG and LCB first. With this method, the table given above will have levers calculated from LCF of the original waterline and the moment sum is the trimming moment. This can then be divided by the MCT for the new draught to produce the change in trim.

The main assumption of this approximation is that the centroid of the parallel-sinkage layer can be said to be at the LCF of WL. This assumption will not introduce too much error for a conventional ship-shape when the change in displacement is fairly small (for example, draught change of a few cm rather than a few tens of cm).

Example 5.1 A box-like vessel 100 m L × 30 m B is operating at an even keel draught of 6 m. At a port the following loading/unloading operations have been carried out:

Calculate the trim and consequently the draughts fwd and aft.

Solution

All items can be calculated directly, since the vessel is box-like. Thus, $\nabla = L \times B \times T$ and the increase in draught due to the added cargo is $\delta T = \frac{\sum m}{L \times B}$

Mass (tonnes)	Loading/unloading	x_g (m from midship)
100	L	10 fwd
50	U	40 aft
200	L	30 aft
50	U	20 fwd

Mass	Lever	Moment
+100	+10	+1000
-50	-40	+2000
+200	-30	-6000
-50	+20	-1000
$\Sigma m = 200$		$\Sigma mx = -4000$

$$\nabla = L \times B \times T \quad \text{and the increase in draught due to the added cargo is}$$

$$\delta T = \frac{\Sigma m}{L \times B}$$

$\nabla_0 = 100 \times 30 \times 6 = 18,000\,\text{m}^3 \quad \Delta_0 = 18,000 \times 1.025 = 18,450\,\text{tonnes}$

$\Delta_1 = 18,450 + 200 = 18,650\,\text{tonnes} \quad \nabla_1 = 18650/1.025 = 18,195.1\,\text{m}^3$

$T_1 = 6 + 200/(100 \times 30 \times 1.025) = 6.065\,\text{m}, \quad \text{trimming moment} = -4000\,\text{tonne-m}$

$I_L = 30 \times 100^3/12 = 2,500,000\,\text{m}^4 \quad BM_L = I_L/\nabla_1 = 137.4\,\text{m}$

$\text{MCT} = 18,650 \times 137.4/(100 \times 100) = 256.25\,\text{tonne-m/cm}$

$\text{trim} = -4000/256.25 = 16\,\text{cm by aft.}$

Since LCF is amidships, trim forward and aft will be identical. Thus,

$$T_{aft} = T_1 + 0.16/2 = 6.145\,\text{m}, \quad \text{and} \quad T_{fwd} = T_1 - 0.16/2 = 5985\,\text{m}$$

Example 5.2 Question

A container ship is operating at a level keel draught of 15 m in sea water with a displacement of 15,000 tonnes. LBP of the ship is 100 m. In this condition the vessel has the following characteristics:

Centre of buoyancy at 10 m aft of midship and 5.5 m above baseline.
VCG at 8 m above baseline.
TPC 35 tonnes/cm.
LCF at 5 m aft of midship
$I_L = 1,351,303\,\text{m}^4$.

(I_L, *TPC and LCF may be assumed to be constant for the range of draughts we are interested in here.*)

The following loading/unloading operations are carried out:

Operation	Mass (tonnes)	From midship	Above baseline (m)
load	150	20 m fwd	5
Unload	220	40 m aft	9
Unload	540	5 m aft	6
Load	325	25 m aft	10
Load	400	12 m aft	7
Load	100	35 m aft	9

Estimate the new values for the following items:

LCG
LCB
VCG
VCB
draughts forward and aft

Solution
There are a number of ways of starting this, but one simple way is as follows:

Operation	Mass	x_i	$m_i x_i$	z_i	$m_i z_i$
Load	150	20	3000	5	750
Unload	-220	-40	8800	9	-1980
Unload	-540	-5	2700	6	-3240
Load	325	-25	8125	10	3250
Load	400	-12	-4800	7	2800
Load	100	-25	-3500	9	900
Σ	215		14,325		2480

New displacement $= 15,000 + 215 = 15,215$ tonnes
Since originally LCG must have been the same as LCB,
new LCG $= \frac{15,000 \times (-10) + 14,325}{15,215} = -8.917\,\text{m}$
(or 8.917 m aft of midship)
New LCB $= \frac{15,000 \times (-10) + 215 \times (-5)}{15,215} = -9.929\,\text{m}$
(or 9.929 m aft of midship)

New VCG $= \frac{15{,}000 \times 8 + 2480}{15{,}215} = 8.050$ m

Change in draught $= 215/35 = 6.1$ cm

New draught $= 15.061$ m

Height of c.g. of parallel sinkage layer $= 15.031$ m

$$\therefore \text{new VCB} = \frac{15{,}000 \times 5.5 + 215 \times 15.031}{15{,}215} = 5.635 \text{ m}$$

$$MCT \; 1\,\text{cm} = \frac{\Delta GM_L}{100\,L}$$

$$GM_L = KB + BM_L - KG$$

$$= 5.635 + I_L\big/\nabla - 8.05 = 88.619 \text{ m}$$

$$\therefore MCT \; 1\,\text{cm} = \frac{15{,}215 \times 88.619}{100 \times 100} = 134.83 \text{ tonnes-m/cm}$$

Trimming moment $= 15{,}215 \times (9.925 - 8.917) = 15{,}336.7$ tonnes-m

Trim $= 15{,}336.7/134.83 = 113.7$ cm $= 1.137$ m

Trim aft $= \frac{-45}{100} \times 1.137 = -0.512$ m

Trim fwd $= 1.137 - 0.512 = 0.625$ m

Therefore, draught fwd $= 15 + 0.625 = 15.625$ m

draught aft $= 15 - 0.512 = 14.488$ m

Key Points

- trim $=$ trim fwd $+$ trim aft

- $BM_L = \frac{I_L}{\Delta}$

- $GM_L = KB + BM_L - KG$

- $MCT = \frac{\Delta GM_L}{100L} \approx \frac{\Delta BM_L}{100L}$ tonne-m/cm

- Change in trim $=$ (trimming moment)/MCT cm

Exercises

1. A wall-sided barge of 50 m L × 20 m B has a constant waterplane shown below.

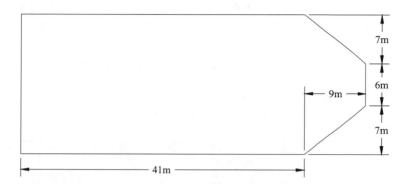

The barge is operating in seawater at an even keel draught of 3 m. Assume that the centre of gravity of the vessel is at 4 m above keel at all loading conditions. (*The events described in this question occur sequentially in the order shown.*)

(a) The vessel sails up-river into a fresh water lake.
   - Compute its new draught.
   - Does the barge change trim? (Explain briefly the reason.) If so, what is the change in trim?
   - Compute the position of LCF and hence I_L.

(b) She then discharges some cargo as follows:
   50 tonnes from 20 m aft of midship.
   The following are loaded:
   100 tonnes at 15 m aft of midship.
   10 tonnes at 20 m fwd of midship.
   - Compute the new mass and volume displacement.
   - Calculate BM_L, GM_L and MCT.
   - Estimate the new draughts fwd and aft.

(c) The barge then returns down-river to sea. Does the trim change? Explain clearly why you think so.

2. A ship, 150 m L × 25 m B is operating in sea water at a draught of 6 m even keel. In this condition the half-breadths and the cross sectional area (CSA) for the waterplane are as follows:

Station	0	1	2	3	4	5	6	7	8	9	10
Half-breadths (m)	3.80	10.00	12.00	12.50	12.50	12.50	12.50	12.50	11.00	7.20	0.00
CSA (m²)	72.7	117.1	140.0	140.0	140.0	140.0	140.0	136.0	114.0	52.30	0.00

(a) Calculate TPC and the position of LCF
(b) Calculate the volume displacement and the position of LCB
(c) Calculate I_L and I_T
(d) Calculate BM_L and GM_L (Assume KG = 8 m and KB = 3.3 m)
(e) Additional cargo of 200 tonnes is added with its centre of gravity at 32 m
     aft of midship. Estimate the draughts forward and aft. Clearly state any
     assumptions you make.

3. A box-like ship, 100 mL × 25 mB × 10 mD is operating in sea water at an even
    keel draught of 6 m. The centre of gravity of the ship is at 10 m above keel at
    this condition.
    At a port, the following loading/unloading operation is carried out:

Mass (tonnes)	Operation (L/U)	c.g. from midship (m) (*positive fwd of midship*)
200	L	32
100	L	−18
300	U	20
1000	L	−30
2000	U	−10
2000	L	−20
500	L	40

(a) Compute the new even keel draught
(b) Compute BM_L and, thus, MCT
(c) Estimate the trimming moment and, hence the total trim, draught fwd and
     draught aft.

4. A wall-sided vessel with the constant waterplane shown below is operating in
    seawater at an even keel draught of 4 m.

(a) Calculate C_m, C_p and C_b.
(b) Calculate TPC.
(c) Estimate the position of LCF.
(d) Calculate I_T and I_L.
(e) Calculate BM_L and hence MCT.
(f) A cargo of 300 tonnes was moved aft by 10 m. Estimate the new draughts
     aft and fwd.

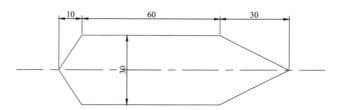

5. The following data, which may be assumed to be constant except for T_{aft}, are given for a vessel of 100 m length as follows:

   TPC = 18.0 tonnes/cm, MCT = 160 tonne-m/cm
   LCF is amidships, original T_{aft} = 10.0 m
   After ballast is discharged from a tank with c.g. at 52 m aft of midship, the vessel is found to float at a mean draught of 9.5 m and to have a trim of 50 cm by the stern. Calculate the amount of ballast discharged and the original draught forward.

6. A rectangular vessel 120 m L × 30 m B × 7 m T has VCG at 6 m above keel and LCG amidships. The ship goes aground at the forward end, and then the tide falls by 0.5 m.
   Calculate the draught aft of the ship in this condition. In order to effect refloating of the vessel several pieces of cargo weighing total 1000 tonnes with their centroid at 30 m fwd of midship are to be moved. Find the minimum horizontal distance by which they have to be moved.

Chapter 6
Statical Stability at Small Heel Angles

6.1 Introduction

All ships should have adequate stability in all their operating conditions. This state-ment sounds almost too obvious and simple to deserve a serious consideration, but it raises two important questions: firstly, what is stability; and secondly how do we define 'adequate'? We cannot even begin to answer the second until we have answered the former.

The word 'stability' can mean different things to different people: for example, mathematicians talk about the stability of solutions, dynamicists discuss the stability of a system, and chemists worry about the stability of a compound they are working on, while many people know roughly what is meant by economic, political or social stability. Although stability is used in Naval Architecture in a variety of contexts as well, for example in 'directional stability', the unqualified word 'stability' in normal usage has a specific meaning, reflecting our preoccupation in safety and keeping a ship upright, implying that achieving this aim cannot be taken for granted.

A more precise definition of ship stability may be 'the ability of the ship to return to the normal upright condition, when disturbed from this attitude, without endangering itself or the cargo and human life it carries'. The keyword here is 'return', and this means that the disturbance from the normal upright condition (equilibrium), whatever caused it, will automatically generate an opposing force or moment to return it to the original equilibrium. The nature of these 'restoring' forces and moments will become clearer if we examine the cases of disturbances in different directions.

First of all, it is easy to see that the disturbances on the horizontal plane, i.e. displacements along x- and y-axes and rotation about z-axis (surge, sway and yaw respectively) will not generate any restoring force or moment, as the system in these cases is in complete neutral equilibrium. Displacement along z-axis (heave) can be relied on to produce restoring force, provided the waterplane area is non-zero. In the case of normal ships this, therefore, is no problem, except that there should be sufficient freeboard so that the ship faces less chance of being swamped. In any case,

© Springer Nature Singapore Pte Ltd. 2019
B. S. Lee, *Hydrostatics and Stability of Marine Vehicles*, Springer Series on Naval
Architecture, Marine Engineering, Shipbuilding and Shipping 7,
https://doi.org/10.1007/978-981-13-2682-0_6

the static displacement in heave direction is straight forward and does not require complicated theory to explain the phenomenon.

On the other hand, rotational disturbances about x- and y-axes (roll and pitch, or their static equivalent of heel and trim) are very different indeed. One knows, instinctively, that the disturbance about the y-axis (i.e. trim) is no problem, because longitudinal stability is so high for conventional ships. Nevertheless this is an important factor in determining the performance of ship propulsion and this has been discussed in Chap. 5.

This leaves the stability in the transverse direction as the really important issue as we cannot be sure that the ship will generate sufficient moment to return to upright condition when heeled and how much heel the ship can suffer without completely capsizing. Indeed the word capsize normally implies excessive heeling.

We shall, therefore, examine the transverse stability carefully, with particular emphasis for the cases where the watertight integrity of the ship is intact (intact stability). When we are completely at ease with this, we shall examine the approaches normally used for studying the stability when the ship is damaged and the normally watertight skin of the ship is breached.

6.2   Basic Principles of Stability

Before we start examining the nature of the restoring moment (or more commonly known as 'righting moment'), we need to make some fundamental assumptions as follows:

(a)   the ship is floating in calm water,
(b)   the ship is stationary, and
(c)   the ship is rigid, i.e. it maintains its original shape whatever attitude it might assume.

Some of these assumptions can be removed either partially or wholly, and we shall deal with some of them in later chapters.

Righting Moment

We shall now have a look at the mechanism through which the righting moment is generated when an arbitrary object is inclined from an equilibrium position. Consider an object floating in water at equilibrium. The weight of the object is W and its centroid of mass is at G. Since the object is in equilibrium, the underwater volume of the object will be $\frac{W}{\rho g}$, where ρ is the mass density of the water and g is the gravitational acceleration. Furthermore the zero sum of moments dictates that the centroid of the underwater volume, B, be vertically under G as shown in Fig. 6.1a.

If this object is then inclined by an angle ϕ, as shown in Fig. 6.1b, without altering the displacement, the centroid of the underwater volume will move to a new location B_1, while the mass centroid of the object remains at G. It is easy to see that, except in very special circumstances, B_1 will in general swing out so that it is no longer

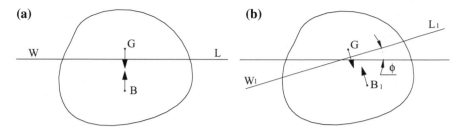

Fig. 6.1 **a** Object in equilibrium. **b** After heeling

vertically under the point G. The sum of the forces is zero but a moment is created, and, therefore, the object is no longer in equilibrium. The lever of this moment is the distance between the two vertical lines through G and B_1 and the moment may be in either direction. You may recall from Chap. 5 that in ship stability righting moment often refers to mass moment instead of force moment. We shall use this convention unless there is a clear need to use force moment. The magnitude of the righting moment, therefore, is simply the product of the mass W/g and the lever. This lever is known as the '*righting lever*' or '*righting arm*'.

Note that a 'heeling moment' tends to increase the angle of inclination. Therefore, a positive heeling moment will be in the same direction as a negative righting moment. If we divide the heeling moment by the mass of the object, we obtain '*heeling lever*' or '*heeling arm*'.

6.3 Symbols and Terminology

When stability is discussed in naval architecture we need to define the meaning of certain terminology and symbols. While some authors use their own systems, the most commonly used symbols and terminology are illustrated in Fig. 6.2.

6.4 Metacentric Height

It is convenient to assign a positive or negative sign to the angle of heel, and throughout this book we shall use the convention that heel to starboard is positive and heel to port is negative (viewed from the stern). For example, a heel angle of 10° means 10° heel to starboard and $-10°$ denotes 10° to port.

Consider a vessel floating in equilibrium at WL is heeled to starboard by an angle ϕ so that the new waterline can be represented by W_1L_1. Note that at this waterline the displacement should be the same as the original one, and for a small angle of heel W_1L_1 can be assumed to pass through point O of Fig. 6.2. The CG of the vessel is

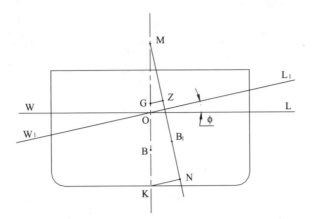

WL: upright waterline
W_1L_1: heeled waterline
G: centre of gravity of the vessel
B: centre of buoyancy (upright)
B1: centre of buoyancy (heeled)
K: keel (or the point at which the baseline crosses the centreline)
M: (initial transverse) metacentre (written M_T if there is danger of confusion)
ϕ: angle of heel
KB: height of V.C.B.
KG: height of V.C.G.
GM: (transverse) metacentric height (sometimes written GM_T)
BM: (transverse) metacentric radius (sometimes written BM_T)
GZ: righting lever (or righting arm)

Fig. 6.2 Definition of symbols and terms

still at G (assumed to be on the centreline of the vessel), while CB will move to B_1. A righting moment is generated as indicated by the two forces equal in magnitude, but opposite in direction acting on two vertical lines separated by a distance GZ. Thus,

$$M_R = GZ \times \Delta$$

GZ is the righting lever.
For small ϕ, it can also be shown that

$$GZ = GM \sin \phi \approx GM \phi$$

Remember that ϕ is in radians in the above expression.

Now that we have the stability quantified, albeit in a very rudimentary way, we need to examine what the point M signifies and how we can calculate GM. Point M, or the **metacentre**, is formally defined as the point of intersection of the two vertical lines passing through the centres of buoyancy at two very close angles of inclination. It is easy from this definition to see why the length BM is called '**metacentric**

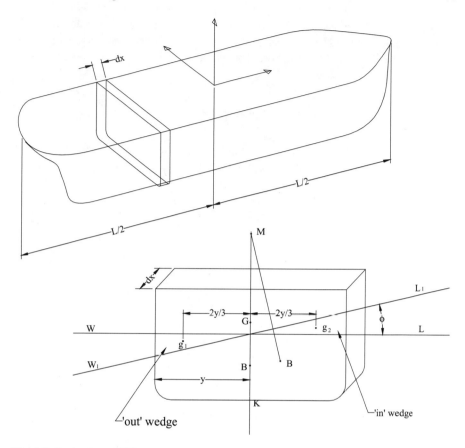

Fig. 6.3 Derivation of GM

radius'. The metacentre at the heel angle of zero is called the initial metacentre, or more often simply as the metacentre. The metacentres at non-zero angles of heel are known as **pro-metacentres**. It is observed from geometry that

$$GM = BM + KB - KG$$

Therefore, if we can calculate BM and KB, and if we know KG, we can obtain GM. The method of calculating KB and KG has already been discussed in earlier chapters, and thus our problem is reduced to determining BM.

Consider Fig. 6.3, which shows the whole ship and we attempt now to calculate GM (called **metacentric height**, or more correctly *initial* **metacentric height**). The displacement in the heeled state is the same as the initial displacement. Therefore, the volume of the emerged wedge ('out' wedge), v_1, should be the same as that of the immersed wedge ('in' wedge), v_2. In other words, $v_1 = v_2 = v$, say.

For very small angles of heel, the in- and out-wedges can be considered to be triangular in section and so the local centre of gravity of the wedges is $\frac{2}{3}y$ away from the centreline in transverse direction. The moment of transfer of buoyancy for this thin strip of thickness dx is

$$2 \times \frac{2}{3}y \left(\frac{1}{2}y \cdot y \tan \phi \right) dx = \frac{2}{3}y^3 \tan \phi dx$$

If we now integrate this moment along the whole length of the ship (i.e., from $x = -L_a$ to $x = L_f$), we obtain the total moment of transfer for the whole ship.

$$\int_{-L_a}^{L_f} \frac{2}{3}y^3 \tan \phi dx = \frac{2}{3}\tan \phi \int_{-L_a}^{L_f} y^3 dx$$

We can relate this moment to the transverse movement of the centre of buoyancy which we will call h.

From the moment theory we know that

$$h = \frac{\frac{2}{3}\tan \phi \int_{-L_a}^{L_f} y^3 dx}{\nabla}$$

On the other hand, for small ϕ

$$h \approx BM \tan \phi$$

Therefore

$$BM = \frac{2}{3} \int_{-L_a}^{L_f} y^3 dx / \nabla$$

Note that y is half breadth of the waterline, and therefore is a function of x.

The integral $\frac{2}{3} \int_{-L_a}^{L_f} y^3 dx$ happens to be the second moment of the waterplane about the centreline, or I_T (transverse second moment of the waterplane).

6.5 Implications of GM

Having found a way of calculating GM, we now need to explore its implications a little before we go further into stability.

(i) **GM as a stability parameter**
As we have seen, the righting moment of a vessel when heeled to a small angle ϕ is

$$\text{Righting moment} = \Delta GM \sin \phi \approx \Delta GM \phi$$

In other words GM is a direct measure of stability at small angles of heel. The reason why people prefer GM as a parameter to GZ, for example, is that GM does not vary whilst GZ is a function of ϕ. Thus, GM is a parameter showing the initial stability characteristics of the vessel.

For reasons which will become apparent later on, at larger angles GM often gives misleading information about stability and consequently it is unwise to use it alone to assess the stability of a vessel.

(ii) **GM as a roll motion parameter**
Although the full implication of GM in terms of roll motion characteristics have to be explored later, it is worth noting at this stage that the natural period of small amplitude roll motion of a vessel is a function of GM, i.e.

$$T_n = \frac{2\pi k}{\sqrt{GM \cdot g}}$$

where

T_n natural roll period
K radius of gyration (often about 0.42 B)
G gravitational acceleration.

It can be seen, therefore, that a vessel with high initial GM (i.e. a 'stiff' ship) will have short natural roll period, and a 'soft' ship (low GM) will have long natural period. A stiff vessel will roll relatively rapidly in a jerky fashion which may be detrimental to the comfort of passengers and crew. Note the use of terms analogous to the characteristics of a spring.

6.6   Methods of Influencing GM

The easiest way of changing GM of an existing vessel is by using ballast, often in solid form, to lower the centre of gravity. This often is an expensive, albeit sometimes necessary, way of increasing GM, as it reduces the carrying capacity of the ship whilst slowing the ship down. At the design stage, however, any perceived shortcomings in intact transverse stability can be remedied by increasing the second moment of the waterplane area, either by increasing the breadth of the ship, where it is possible, or changing the shape of the waterplane. It must be remembered that the waterplane is often governed by the requirements of speed or resistance, and therefore the designers are not entirely free to determine its shape for stability alone.

(a)

(b)

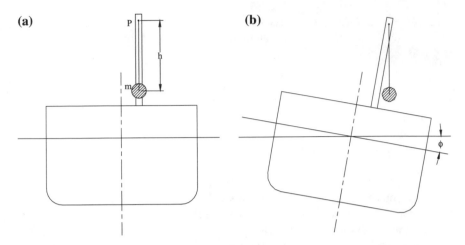

Fig. 6.4 **a** Suspended weight in upright condition. **b** Suspended weight when heeled

6.7 Factors Affecting GM

The GM calculated as above may not be the effective GM for the ship, depending on its conditions. There are a number of factors which need to be taken into account when trying to arrive at an effective GM and the two typical cases are suspended weight and free surface effects both of which are due to movement of the load on board.

Suspended Weight
Consider a vessel with a weight of mass m suspended from point P as shown in Fig. 6.4a. First, if we assume that the weight is tied to the post so that it does not move, then the righting moment of the vessel for a small angle of inclination ϕ is

$$RM = \Delta \cdot GM \sin \phi$$

If the weight is gently released in this heeled condition, the weight will move to a position vertically below the point P as shown in Fig. 6.4b.
Then the movement of the centre of gravity of the weight is

$$gg_1 = h \sin \phi$$

we can now approach the problem in two ways.

(i) The heeling moment due to the shift of the weight centre of gravity is $m \cdot h \cdot \sin \phi$

Therefore the net righting moment

$$= \Delta GM \sin \phi - mh \sin \phi = \Delta \left(GM - \frac{mh}{\Delta} \right) \sin \phi$$

In other words, GM is effectively reduced by mh/Δ.

(ii) The shift of the vessel centre of gravity due to the shift of the weight centre of gravity is $GG_1 = \frac{mh \sin \phi}{\Delta}$

Therefore, the net righting moment

$$= \Delta(GM \sin \phi - GG_1 \cos \phi) \approx \Delta \left(GM - \frac{mh}{\Delta} \right) \sin \phi,$$

since $\cos \phi \approx 1$ for $\phi \ll 1$.

This is the same expression as obtained in (i).

The apparent reduction in GM of $\frac{mh}{\Delta}$ is equivalent to an increase in VCG by the same amount and this signifies that, when evaluating the VCG of a vessel, any suspended weight free to swing should be considered to be situated at the points of suspension.

Free Surface

The free surface effects are by far the most important factor which influences the magnitude of initial GM. These effects occur when there is any container with liquid content having free surface, i.e. the surface of the liquid free to maintain horizontal position whatever the ship's attitude (termed a 'slack' tank).

Consider a vessel with a tank partially filled with liquid of density ρ_T. The density of the water the ship is floating on is ρ_W and the volume of the fluid is v.

The righting moment of the vessel if the surface is not allowed to move (e.g. by assuming a rigid surface film) at a small angle of heel ϕ is $\Delta \cdot GM \sin \phi$. This righting moment is often called 'solid' righting moment, and GM is called solid GM and sometimes denoted as GM_s.

Now the thin film on the surface is removed and the liquid is allowed to move. For the sake of clarity we shall consider a rectangular tank of breadth $2b$ and length l as shown in Fig. 6.5. The point m denotes the intersection of the line at right angles to WL from g and the line at right angles to W_1L_1 from g_1, where g is the original centre of gravity and g_1 is the centre of gravity of the liquid at the heeled condition.

$$\text{The moment of volume transfer} = \frac{2}{3} b \cdot b \cdot b \tan \phi \cdot l = \frac{2}{3} b^3 l \tan \phi$$

Therefore,

$$gg_1 = gm \tan \phi = \frac{2}{3} lb^3 \tan \phi / v = \frac{i_T}{v} \tan \phi.$$

Again there are two approaches.

Fig. 6.5 Vessel with a
'slack' tank

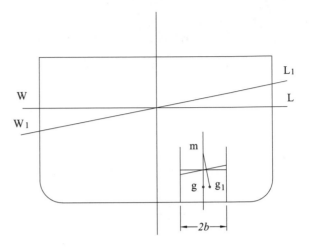

(i) The c.g. of the liquid will move from g to g_1 and therefore the C.G. of the vessel moves to G_1.

$$GG_1 = \frac{w \cdot gg_1}{\Delta} = \frac{w \cdot gm \tan \phi}{\Delta} = \frac{w}{\Delta} \frac{i_T}{v} \tan \phi$$

$$= \frac{i_T}{\Delta} \frac{v \rho_T}{v} \tan \phi = \frac{i_T \rho_T}{\Delta} \tan \phi$$

where

w is the liquid mass $= v\rho_T$
v is the liquid volume
i_T is the local transverse second moment of the free surface area.

Thus, the net righting moment

$$= \Delta(GZ - GG_1 \cos \phi) = \Delta \left(GM - \frac{i_T \rho_T}{\Delta} \right) \sin \phi$$

$$= \Delta(GM - GG_2) \sin \phi, \text{ say.}$$

This item GG$_2$, which is equal to $\frac{i_T \rho_T}{\Delta}$, is called the 'free surface effect' and the action of subtracting the free surface effect from GM solid is called 'free surface correction'. The term $GM - \frac{i_T \rho_T}{\Delta}$ is known as the GM fluid and denoted GM$_f$. It can be seen that the effect of the free surface is to increase the VCG effectively by this amount. In other words, recalling a similar situation with suspended weights, the c.g. of the liquid appears to be at the point m and not at g.

(ii) Heeling moment due to the shift of the liquid c.g. is

$$w \cdot gg_1 \cos \phi = w \cdot gm \tan \phi \cos \phi = w \cdot gm \sin \phi$$

Therefore,

$$\text{the net righting moment is} = \Delta GM \sin \phi - w \cdot gm \sin \phi$$
$$= \Delta \left(GM - \frac{w \cdot gm}{\Delta} \right) \sin \phi$$
$$= \Delta \left(GM - \frac{i_T \rho_T}{\Delta} \right) \sin \phi$$

as obtained in (i).

It is not difficult to see a certain similarity between free surface effect and suspended weight effect.

For suspended weight

$$\text{net righting moment} = \Delta \left(GM - \frac{mh}{\Delta} \right) \sin \phi$$

For free surface

$$\text{net righting moment} = \Delta \left(GM - \frac{w \cdot gm}{\Delta} \right) \sin \phi$$

6.8 Inclining Experiment

It can be seen from the above that in order to get an accurate idea of stability we need to know the accurate position of the VCG of the vessel. The VCG is estimated at the design stage, but this estimation is not sufficiently certain to rely upon. Furthermore, the VCG can change after being in service for a while. Therefore, all ships have to undergo a measurement process known as inclining experiment or test after launch and at regular intervals thereafter.

Since directly measuring KG is not possible, GM is estimated by imposing a known heeling moment and measuring the resulting heel angle. Since KM can be calculated from the hydrostatics, KG can be estimated from this.

A clear windless day and a site with minimum current is selected for this experiment. All mooring lines are slackened off and all non-essential personnel are taken off the ship. There should be no suspended weight or slack tank, but, if it is not possible to ensure this, the details of these should be taken so that they can be taken into account. The draught marks around the ship are read off so that the displacement, VCB and LCB can be calculated. Any items which are not a normal part of the ship should also be removed or, if it is not possible, taken a careful note of.

A number of similar weights, w, are then moved across the deck through a known distance, h, so as to incline the vessel to port and starboard in turn by a small angle. The angles of inclination are carefully measured at each stage by either an inclinometer or a long pendulum with a plumb bob in a bucket of oil to dampen its movement. If identical inclining weights are used, then an average change in inclination after each movement of the weight is obtained, and then

$$GG_1 = \frac{wh}{\Delta}$$

$$\text{but } GG_1 = GM_T \tan \phi$$

$$\text{Therefore } GM_T = \frac{wh}{\Delta \tan \phi}$$

If a pendulum is used, $\tan \phi$ can be obtained by dividing the horizontal movement of the pendulum by its length. Having obtained KG ($KG = KB + BM_T - GM_T$), it is relatively simple to undertake corrections for removing non-lightship items, as we need to keep a careful track of KG for the light ship condition.

Key Points

- $BM = \frac{I}{\nabla}$
- Righting arm GZ = righting moment/Δ
- Heeling lever = heeling moment/Δ
- $GM = BM + KB - KG$
- $GZ \approx GM \sin \phi$ for $\phi \ll 1$
- Free surface effect $= \frac{i_T \rho_T}{\Delta}$
- From inclining test $GM = \frac{wh}{\Delta \tan \phi}$

Exercises

1. A box-shaped barge of $50m^L \times 10m^B \times 3m^T$ is floating in calm sea.

    (a) If the VCG of the barge is 4 m above the baseline, find the initial metacentric height.
    (b) How high the centre of gravity can be raised without causing GM to become negative.
    (c) With KG = 4 m, increase length, breadth and draught by 10% one at a time and compute GM for each case. Comment on the results.

2. A box-like vessel of $100m^L \times 20m^B \times 10m^D$ is operating in sea water at a draught of 6 m. In this condition KG = 7 m. It has a box-shaped oil tank measuring $20m^L \times 20m^B \times 3m^D$. The centreline of the tank is at midship and the tank is initially full. Assume that the tank rests on the bottom of the vessel.

(a) Find the initial GM.

(b) If half of the oil (specific gravity 0.9) is pumped out, what is GM_f.

(c) Instead of pumping the oil out as in (b), one third of the original full tank content is pumped into a settling tank measuring $20m^L \times 10m^B \times 3m^D$ and situated right above the oil tank. Find GM_f for this situation.

3. A box-like vessel, $180m^L \times 31.5m^B$, floats in sea water at a level keel draught of 12 m and in this condition the value of GM is 0.891 m. A 'tween deck space, $30\ m^L \times 31.5\ m^B$, is accidentally flooded with sea water to a depth of 1.35 m. The deck on which the water is lying is level and is 12.6 m above the baseline of the vessel. Estimate the effective GM (or GM_f) of the vessel.

4. A box-like barge, $80m^L \times 15m^B \times 10m^D$, is operating in sea water at an even keel draught of 3 m. When some cargo of mass 200 tonnes was moved horizontally in the transverse direction by 3.2 m, the barge heeled by $3°$.

(a) Estimate GM_T and BM_T, and hence the position of the VCG.

(b) Having returned the moved cargo to its original position, it is now decided to load the barge with some containerised cargo of mass 2000 tonnes on the deck. The centre of gravity of the additional cargo when loaded will be at 11.5 m above keel. Estimate the transverse metacentric height.

(c) If the answer to (b) is a negative number, ignore the operation in (b). Instead estimate what is the allowable cargo mass which can be loaded on the deck, if the minimum allowable GM_T is 1 m? (Ignore trim and assume the c.g. of the cargo at 11.5 m above keel.)

5. A semi-submersible comprises two rectangular cross-section pontoons, each 90 m long, 11 m wide and 7 m deep. The deck is supported by four cylindrical columns each 9.5 m in diameter, the centrelines being spaced 70 m apart longitudinally and 60 m apart transversely. The vessel floats at a draught of 23 m in sea water and has a vertical centre of gravity 18.5 m above the keel. Calculate the transverse metacentric height.
A mass of 800 tonnes is added on board at a position 35 m above the keel. Calculate the new values of draught, vertical centre of gravity and the transverse metacentric height.

6. A box-like crane barge, $100m^L \times 25m^B \times 10m^D$, is operating in sea water at an even keel draught of 6 m. The centre of gravity of the vessel is at 10 m above keel.

(a) Compute GM_T.

(b) The crane on board picks up a piece of cargo of mass 500 tonnes from the deck. The crane boom tip is at 20 m above the deck. Estimate the new virtual KG and hence GM_T.

(c) One possible way of increasing the stability is placing some heavy substance, such as concrete ballast, at the lower part of the barge. Suppose that it is proposed to place a solid ballast of 2000 tonnes at 1.5 m above keel to achieve this. Calculate the GM_T when this mass has been introduced and the operation in (b) takes place.

   (d)   Do you think this ballast is an efficient way of achieving the required result?

7.   A box-like barge, 80 m ^{L} × 20 m ^{B}, is operating upright in sea water at an even keel draught of 6 m. When a piece of deck cargo of mass 40 tonnes was moved to starboard by 10 m, the barge heeled to starboard by 2°.

   (a)   Estimate the VCG of the vessel.

   (b)   While the ship is heeled 2° to starboard, some cargo of total mass 100 tonnes with its c.g. at 6 m from centreline to port and 4 m above the keel is to be unloaded. Estimate the resulting heel angle to be expected.

Chapter 7
Statical Stability at Large Heel Angles

7.1 Introduction

The stability characteristics of a vessel at large angles of heel are in general very much different from those of initial stability or stability at very small angles of heel. It may be recalled that stability of a vessel can be quantified in terms of righting moment or, for a given displacement, righting arm. Moreover, the righting arm could be represented by using the initial metacentric height in the form.

$$GZ = GM \ \sin \ \phi$$

The most essential assumption in the derivation of this expression was that $\phi \ll 1$. In other words, at very small angles of heel we can take the initial metacentric height as the stability parameter.

However, as ϕ increases, this relationship becomes less valid and beyond the heel angle of about 10° GM loses its meaning as the only stability parameter and, indeed, it can often be misleading. Therefore, it is now necessary to go back to the righting arm for a meaningful indication of stability. This section examines various ways of calculating the righting levers at large angles of heel, but first we shall discuss what happens to the by now familiar term of metacentre at large angles.

When dealing with initial stability, $BM = I_T / \nabla$ was developed with the assumption that the centre of buoyancy moves horizontally, i.e. that the transfer of wedges was horizontal. Obviously this cannot be true at large angles and the centre of buoyancy moves out along a curve called the **isoval** or **metacentric involute**.

When the ship is upright, this curve is perpendicular to the centreline of the ship and the radius of curvature ρ is $B_0M = I_T/\nabla$. The verticals through the heeled centres of buoyancy do not normally intersect at M and any two verticals very small angles apart will have a corresponding metacentre—the pro-metacentre for that angle. The curve along which the pro-metacentres track is known as the **metacentric evolute**.

© Springer Nature Singapore Pte Ltd. 2019
B. S. Lee, *Hydrostatics and Stability of Marine Vehicles*, Springer Series on Naval Architecture, Marine Engineering, Shipbuilding and Shipping 7,
https://doi.org/10.1007/978-981-13-2682-0_7

Fig. 7.1 Cross section of a
ship heeled to a large angle

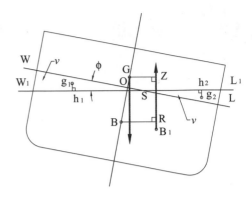

7.2   Attwood's Formula (1796)

Consider a vessel heeled to a large angle ϕ without any change of displacement as shown in Fig. 7.1. Then the volume of the immersed wedge LSL_1 must equal that of the emerged wedge WSW_1, but the inclined waterline W_1L_1 will not necessarily pass through the point of intersection between the original waterline WL and the ship's centreline. Let the volume of the wedges be v with centroids at g_1 and g_2, and the volume displacement be ∇.

Horizontal moment of buoyancy transfer $= v \cdot \overline{h_1 h_2} = \nabla BR$
Therefore, $\nabla BR = v \cdot \overline{h_1 h_2}$
Also from the figure $BR = GZ + BG \ \sin \ \phi$
Therefore, $GZ = v \cdot \overline{h_1 h_2}/\nabla \ - \ BG \ \sin \ \phi$
This is known as Attwood's formula, but in a way it is incomplete, because it does not say how to obtain $v \cdot \overline{h_1 h_2}$. This was done by Barnes, for example (see below).

7.3   Barnes's Method

Consider a ship shown in Fig. 7.2 with the original volume displacement of ∇. As explained above, the heeled waterline W_2L_2 does not necessarily pass through the point O on the centreline, because the volume of the emerged wedge has to be the same as the immersed wedge. On the face of it the exact location of W_2L_2 must be found first before Attwood's formula can be applied. However, a method was found by Barnes by which the righting arm can be calculated without having to do this.

To begin with we consider the waterline W_1L_1 which is parallel to the true heeled waterline W_2L_2 but passing through the point O. This is easily determined as the heel angle is already known. The volume displacement for this waterline (∇_1) is then calculated and compared to the original volume. The real waterline W_2L_2 will be

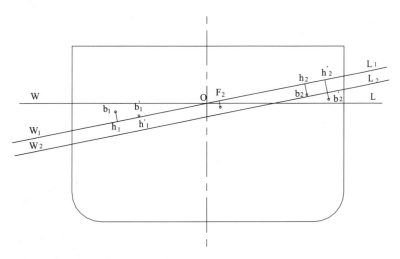

W_2L_2 is the true heeled waterline

v_1 = volume of emerged wedge between WL and W_1L_1

v_2 = volume of immersed wedge between WL and W_1L_1

v = volume of the emerged and immersed wedges between WL and W_2L_2

volume of the layer between W_1L_1 and $W_2L_2 = v_2 - v_1$

b_1', b_2': centroids of v on each side

b_1, b_2: centroids of v_1 and v_2 respectively

h_1, h_2: projections of b_1 and b_2 onto W_1L_1 respectively

h_1', h_2': projections of b_1' and b_2' onto W_1L_1 respectively

F_2 is the projection of centroid of the layer $(v_2 - v_1)$ onto W_1L_1, but since the layer is very small, we can assume it to be identical to the transverse centre of flotation of W_1L_1.

Fig. 7.2 Barnes's method

parallel to W_1L_1 but lower, if $\nabla_1 > \nabla$, and higher, if $\nabla_1 < \nabla$. In other words ∇ can be obtained by subtracting $\delta\nabla = \nabla_1 - \nabla$ from ∇_1. We shall define various points in Fig. 7.2 as follows:

The moment of volume transfer from the emerged wedge to the immersed wedge then is

$$v \cdot \overline{h_1'h_2'} = v_1 \cdot \overline{Oh_1} + v_2 \cdot \overline{Oh_2} - (v_2 - v_1)\,\overline{OF_2}$$

If we evaluate the RHS of this relation, then we can use Attwood's formula. By using radial integration and introducing a dummy variable α as defined in Fig. 7.3,

$$v_1 \cdot \overline{Oh_1} + v_2 \cdot \overline{Oh_2} = \int\limits_{-L_a}^{L_f} \int\limits_0^\phi \left(\frac{1}{2}\,r_1^2\,d\alpha\frac{2r_1}{3} + \frac{1}{2}\,r_2^2\,d\alpha\frac{2r_2}{3}\right)\cos{(\phi - \alpha)}\,dx$$

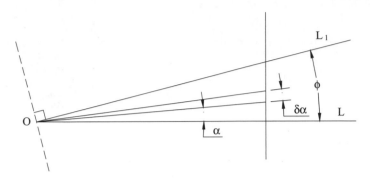

Fig. 7.3 Radial integration used for stability calculation

$$= \int_{-L_a}^{L_f} \int_0^\phi \left(\frac{r_1^3 + r_2^3}{3} \right) \cos(\phi - \alpha) \, d\alpha \cdot dx$$

where r_1 and r_2 are distance between the point O and side hull port and starboard respectively.

v_1 and v_2 can be calculated in a similar manner as follows:

$$v_1 = \int_{-L_a}^{L_f} \int_0^\phi \frac{1}{2} r_1^2 \, d\alpha \cdot dx$$

$$v_2 = \int_{-L_a}^{L_f} \int_0^\phi \frac{1}{2} r_2^2 \, d\alpha \cdot dx$$

Waterplane area $= \int_{-L_a}^{L_f} \left(r_{\phi 1} + r_{\phi 2} \right) dx$

where $r_{\phi 1}$ and $r_{\phi 2}$ are r_1 and r_2 at $\alpha = \phi$ respectively.
First moment of area about the longitudinal axis through O is

$$\int_{-L_a}^{L_f} \left(-\frac{1}{2} r_{\phi 1} + \frac{1}{2} r_{\phi 2} \right) dx$$

Then
$\overline{OF}_2 = $ (first moment)/(area) (note: positive to starboard)

This method can produce the righting arm value at any heeling angle without having to find the true heeled waterline W_2L_2 first.

7.4   Scribanti's Wall-Sided Formula

If the vessel is wall-sided at all points in the length, irrespective of the waterline shape, the areas of immersion and emersion at each section will be equal and opposite right angled triangles and the heeled waterlines for the same displacement will all intersect at the same point on the centreline.

Then by geometry

distance between g_1 and g_2 perpendicular to WL $= \frac{2}{3} y \tan \phi$

area of each triangle $= \frac{1}{2} y^2 \tan \phi$

distance between g_1 and g_2 parallel to WL $= \frac{4}{3} y$

Therefore, moment of transfer of buoyancy parallel to WL

$$= \int_{-L_a}^{L_f} (\tfrac{1}{2} y^2 \tan \phi \cdot \tfrac{4}{3} y) dx$$

$$= \tfrac{2}{3} \tan \phi \int_{-L_a}^{L_f} y^3 \, dx$$

Moment of transfer of buoyancy perpendicular to WL

$$= \int_{-L_a}^{L_f} \left(\tfrac{1}{2} y^2 \tan \phi \cdot \tfrac{2}{3} y \tan \phi \right) dx$$

$$= \tfrac{1}{3} \tan^2 \phi \int_{-L_a}^{L_f} y^3 \, dx$$

But $\frac{2}{3} \int_{-L_a}^{L_f} y^3 \, dx = I_T$, i.e. the transverse moment of inertia of the waterplane.

Therefore

$$\nabla X = I_T \tan \phi \text{ or } X = BM \ \tan \phi$$

$$\nabla Y = \frac{1}{2} I_T \tan^2 \phi \text{ or } Y = \frac{1}{2} BM \tan^2 \phi$$

If the transfer of buoyancy had been purely horizontal (i.e. from B to P), the righting lever would have been

$$GH = GQ \sin \phi$$

But $X = BQ \tan \phi$ and also $BM \ \tan \phi$. Therefore, Q is the initial metacentre M (Fig. 7.4).

The righting lever then is

$$GZ = GH + HZ = GM \ \sin \phi + Y \sin \phi$$

$$= \left(GM + \frac{1}{2} BM \ \tan^2 \phi \right) \sin \phi$$

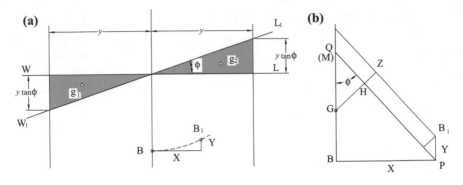

Fig. 7.4 **a** A cross section of a wall-sided ship when heeled; **b** GZ of a wall-sided ship

This formula gives a close approximation of GZ for ships with normal form below the angle of deck edge immersion.

Example 7.1

Consider a log of a constant square section of a homogeneous material with the specific gravity of 0.5 floating in fresh water. Its length is L and the side of the square is a. Let us consider that the log is floating with one of its sides parallel to the water surface.

The volume displacement $V = L \cdot a \cdot \frac{a}{2} = \frac{1}{2} L \cdot a^2$

Transverse second moment of waterplane area $I = \frac{1}{12} L \cdot a^3$

$BM = \frac{2L \cdot a^3}{12 L \cdot a^2} = \frac{a}{6}$

$KB = \frac{a}{4}$

$GM = KB + BM - KG = -\frac{a}{12}$

i.e. the log is unstable when floating on one of its sides. The equilibrium point is when GZ = 0. Based on the wall-sided formula,

$$\tan \phi_E = \sqrt{\frac{-2GM}{BM}} = \sqrt{\frac{2a/12}{a/6}} = 1$$

$$\therefore \phi_E = 45°$$

where ϕ_E is the angle of equilibrium at which the log will float quite happily. In other words the log floats with a corner down.

Fig. 7.5 A typical intact
stability curve

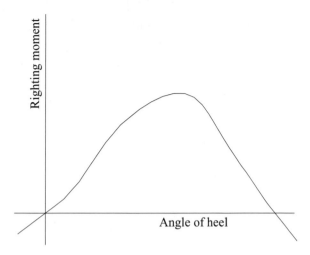

Fig. 7.6 GM and the
maximum GZ

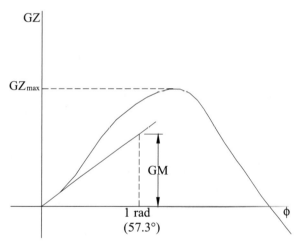

7.5 Curve of Statical Stability

We can learn a great deal about the stability of the vessel by looking at its intact still
water stability curve, and here we shall discuss some of the key features of stability
curves which give us the clues.

A typical stability curve is shown in Fig. 7.5.

Stability curves such as this has the following characteristics:

(a) At small angles of ϕ, $GZ = GM \ \sin \ \phi$. Therefore,
$dGZ/d\phi = GM \ \cos \ \phi = GM$ at $\phi = 0$.
$dGZ/d\phi$ is the gradient of the stability curve and it can be concluded therefore
that the gradient of the righting lever curve at $\phi = 0$ is GM (see Fig. 7.6).

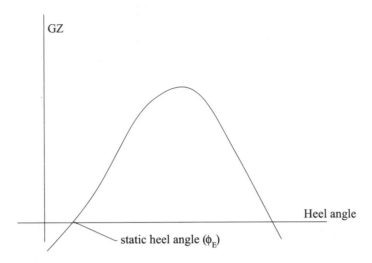

Fig. 7.7 Static heel angle occurs when the centre of gravity if off the centreline

(b) ΔGZ_{max} is the maximum heeling moment the vessel can withstand without capsizing (see Fig. 7.6).

(c) If the centre of gravity of the vessel is not on the centreline, a static heel will occur as shown in Fig. 7.7. This static heel angle is sometimes known as the angle of loll.

(d) If the centre of gravity moves, say, from G to G_1 vertically, the new righting lever at heel angle ϕ is $G_1Z_1 = GZ - GA = GZ - GG_1 \sin \phi$ for all angles. See Fig. 7.8. This is known as sine correction.

(e) If C.G. moves transversely from G to G_1, the new righting lever at angle ϕ is $G_1Z_1 = GZ - GG_1 \cos \phi$ for all angles. See Fig. 7.9. This is known as cosine correction.

(f) The area under the righting moment curve represents work done to heel the vessel or ability of the vessel to absorb energy imparted to it by external forces. Work required to heel from ϕ_1 to ϕ_2 is $\int_{\phi_1}^{\phi_2} \Delta \cdot g \cdot GZ \cdot d\phi$. See Fig. 7.10 (in this diagram the term $\Delta \cdot g$ is omitted, as it is constant). In this case, this work done is identical to the increase in the potential energy of the vessel. If the vessel is then released and it moves from angle ϕ_2 to ϕ_1, this potential energy will be transformed into kinetic energy in the form of increased roll velocity.

(g) 'Range of (positive) stability' is the range of heel angles within which the vessel has positive stability. The final angle at which the stability becomes negative is known as the 'vanishing angle' (see Fig. 7.11)

(h) The stability curves of most merchant vessels are concave at the beginning while occasionally some vessels with pronounced flare, such as sailing yachts, can have convex shape at the beginning. Think about which one is preferable for the same GM.

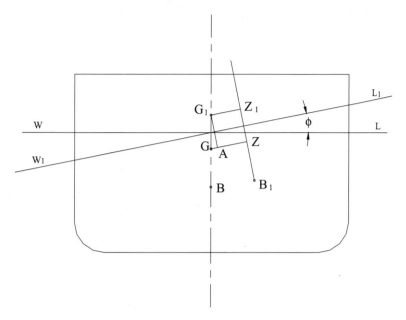

Fig. 7.8 Vertical movement of C.G.

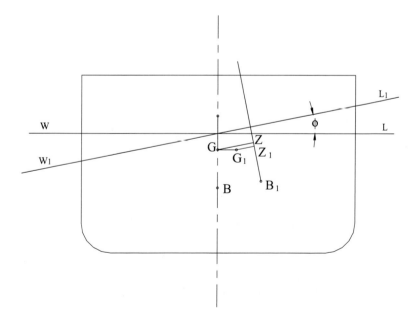

Fig. 7.9 Horizontal movement of C.G.

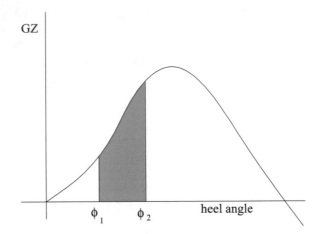

Fig. 7.10 Work done to change heel from ϕ_1 to ϕ_2

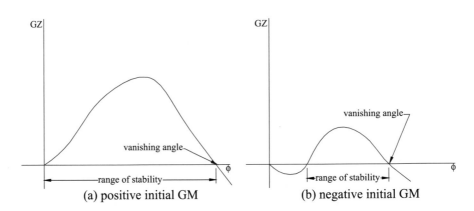

Fig. 7.11 Illustration of vanishing angle (ϕ_v) and range of stability

(i) A negative GM does not necessarily mean the vessel will capsize. It may have a range of positive stability. Note here that the point of positive equilibrium is at the point where the curve crosses the angle-axis upwards as the heel angle increases (see Fig. 7.11b). Remember there will be a similar point at the negative heel angles.

(j) For a symmetric vessel with the C.G. at the centreline, the curve can be expressed as an odd function of heel angle, for example, $GZ(\phi) = a_1\phi + a_3\phi^3 + a_5\phi^5 + \cdots$

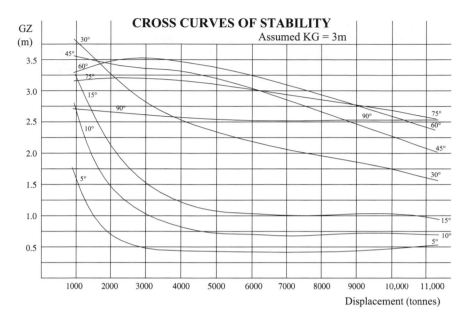

Fig. 7.12 An example of cross curves of stability

7.6 Isocline Method

We have seen that one of major concerns of the Attwood's and Barnes's methods is that the displacement at every heel angle should be identical to the one in the original upright condition. This approach may be called **constant displacement method**. Clearly the major issue here is having to find the correct heeled waterline, even though Barnes's method gets around this problem by approximation. There is another approach which makes no attempt to keep the displacement constant whatever, and it is called **isocline method** (or **constant inclination method**).

The isocline method is an indirect way of obtaining a statical stability curve, and indeed it has an additional advantage that it can cater for any alteration in the loading condition. The result of this method when presented in a graphical form is called **cross curves of stability**, an example of which is shown in Fig. 7.12. It is a convenient and effective method for normal ships but may not be as effective for vessels such as semi-submersibles because the cross curves will then have many discontinuities. A conventional hull form may require information for only about 10 or so draughts and the results are then faired into smooth curves.

The essential procedure of this method can be summarised as follows:

(a) Decide on the angles of heel (often these are 15, 30, 45, 60, 75 and 90°).
(b) For each angle of heel draw several waterlines at least covering the range of displacement of interest.

(c)   Compute the displacement and righting moment about the vertical line through an **assumed** centre of gravity (assumed KG is necessary because KG will change according to the loading condition and hence displacement).

(d)   Plot the righting moments or righting levers on the base of displacement, one curve for each heel angle.

The resulting curves can also be presented in the form of **solid of stability** which is in fact cross curves drawn upon two bases, viz displacement and angle of heel, forming a surface.

Often KG of 0 m is assumed and in this case the cross curves are sometimes called **KN curves**. Whatever value of KG is assumed, it is essential that it is clearly shown in the curves. When using the curves to generate a statical stability curve for a given displacement, it is then necessary to be corrected for the actual KG for that loading condition using the **sine correction** discussed in Sect. 7.5 (d) above. To recap, it will be of the form

$$GZ = (GZ)_0 - G_0G \, \sin \, \phi$$

where

$(GZ)_0$ GZ values obtained from the cross curves
G_0 is the assumed centre of gravity
G_0G is distance between the assumed and real centre of gravity (positive if G is above G_0 and negative if below).

Key Points

- $\frac{dGZ}{d\phi} = GM$ at zero heel angle (the gradient of the GZ curve at $\phi = 0$ is GM)
- Vertical movement of C.G. $G_1Z_1 = GZ - GG_1 \, \sin \, \phi$ (sine correction)
- Transverse movement of C.G. $G_1Z_1 = GZ - GG_1 \, \cos \, \phi$ (cosine correction)
- For wall-sided ships $GZ = \left(GM + \frac{1}{2}BM \, \tan^2 \phi\right) \sin \, \phi$
- Work required to heel from ϕ_1 to ϕ_2 is $\int_{\phi_1}^{\phi_2} \Delta \cdot g \cdot GZ \cdot d\phi$

Excercises

1.   A log of square section with each side 20 cm long is floating in calm fresh water. The specific gravity of the log is 0.5 and the log is homogeneous.

(a)   Find the initial GM
(b)   Find the attitude of stable equilibrium.

2. A box-like barge 100 mL × 20 mB × 10 mD is operating in sea water at an even keel draft of 7 m. The centre of gravity of the vessel is at 6.5 m above keel. Some cragoes of total mass 500 tonnes are transferred athwartship (from port to starboard) by 10 m. Estimate the angle of static heel.

3. A box-like vessel of 100 mL × 30 mB × 13 mD is floating in calm sea water at an even keel draught of 8 m. Its centre of gravity is at 12 m above keel.

   (a) Compute the righting lever at 5, 10, 15 and 18° of heel.
   (b) It is suspected that a part of port side hull, 3 m under the present waterline, is severely corroded. In order to accommodate an inspection, it is desired to bring this area to just above the waterline by heeling the ship to starboard. For this purpose, some weights of total 600 tonnes, which will be supplied by a service vessel, are to be used. These weights can be placed in such a manner that their combined centre of gravity will be at the deck level.

   Estimate the transverse position of the centre of gravity of the weights closest to the centreline which can effect the desired heel.

4. A ship being designed has transverse stability characteristics as represented by the cross curves shown in Fig. 7.12. The ship has a displacement of 6000 tonnes at the operating draught of 6 m. Estimate the maximum allowable KG if the minimum allowable GM is 0.7 m.

5. A cargo ship is operating in sea water at a displacement of 20,000 tonnes. In this operating condition, the statical stability curve of the vessel can be expressed by

$$GZ = 0.7\phi - 0.58\phi^3$$

where

GZ is the righting lever in m
ϕ is the angle of heel in radians.

A cargo of mass 200 tonnes is moved from starboard to port by 10 m horizontally and 5 m vertically upwards. Compute the new statical stability curve and indicate the approximate angle of loll.

Chapter 8
Dynamical Stability

8.1 Introduction

So far the subject of ship stability has been discussed considering that the sea is calm and the ship is stationary, not moving in any mode at all. It is evident that this assumption is wrong and the motion will have to be taken into consideration somehow. It was in mid-1850s when the idea of dynamical stability was introduced in an attempt to relate the stability of a ship to its rolling motion. It is of course clear that most ships which capsize do so after rolling excessively and indeed the capsizing itself is a kind of rolling motion. Therefore, it is quite right that we should relate the rolling motion to stability, although other modes of motion may be ignored for the time being while we consider vessels in their intact condition.

The name dynamical stability, however, is something of a misnomer, basically because it does not deal with the motion in a dynamical way in its true sense. This will be made clear in due course. Nevertheless, the idea has been developed to become the basis of present day **weather criteria** which will be discussed in the next chapter. The essence of this idea is based on energy balance which assumes that there is no energy dissipation during roll motion, in other words the rolling motion of a ship is assumed to be a conservative system, although later development incorporated energy loss through damping terms. Here we shall confine ourselves to the conservative system and examine the concept in some length starting from the roll motion equation.

© Springer Nature Singapore Pte Ltd. 2019
B. S. Lee, *Hydrostatics and Stability of Marine Vehicles*, Springer Series on Naval Architecture, Marine Engineering, Shipbuilding and Shipping 7, https://doi.org/10.1007/978-981-13-2682-0_8

8.2   Basic Equation of Roll Motion

If we assume rolling motion of a ship is a conservative system, we can write the roll equation by using Newton's second law as follows:

$$I_\phi \ddot{\phi} = \sum M$$

where

I_ϕ mass moment of inertia about longitudinal axis through the centre of gravity
$\phi, \ddot{\phi}$ roll angle and acceleration respectively
M   roll moments acting on the ship.

Note that some hydrodynamic moment terms (i.e. added inertia and damping terms) are ignored for the time being.

Let us consider first the case where there is no external heeling moment when the vessel is upright. If we heel the vessel by an angle, say ϕ_1, then there will be a righting moment (which is equivalent to negative heeling moment) equal to $g \Delta GZ(\phi_1)$. Now if we suddenly release the vessel, without imparting any velocity to it, it will roll back to the upright condition gathering speed all the time, and then roll past it to the other extreme losing speed until it stops whence it swings back and so on. In other words it swings about the upright position much like the rhythmic swinging of a pendulum. The roll equation then becomes

$$I_\phi \ddot{\phi} = -g \Delta GZ(\phi)$$

Thus

$$I_\phi \ddot{\phi} + g \Delta GZ(\phi) = 0$$

For small angles of ϕ

$$\Delta GZ(\phi) = \Delta GM \sin \phi \approx \Delta GM \phi$$

Therefore, the roll equation can be linearised to

$$I_\phi \ddot{\phi} + g \Delta GM \phi = 0$$

This equation is similar to that describing free oscillation of an undamped spring-mass system. If the vessel with positive GM was heeled to an angle ϕ_0 and released without imparting initial velocity, it will roll between $+\phi_0$ and $-\phi_0$ with its natural roll period, T_n, which is expressed as

$$T_n = \frac{1}{f_n} = \frac{2\pi}{\omega_n} = \frac{2\pi}{\sqrt{\frac{g \Delta GM}{I_\phi}}}$$

where

f_n natural frequency in Hz or cycles/s
ω_n circular natural frequency in rad/s.

Since $\Delta = \rho\nabla$ and $I_\phi = \rho\nabla k^2$,
where

ρ is the mass density of water
k is the radius of gyration (approximately $0.4B$ for normal ships)
g is the gravitational acceleration,

$$T_n = \frac{2\pi}{\sqrt{\frac{\rho g \nabla GM}{\rho\nabla k^2}}} = \frac{2\pi k}{\sqrt{gGM}}$$

Strictly speaking, this is incorrect because, as you will recall, we did ignore the 'added inertia' and 'damping' terms. However, in rolling motion of most conventional round-bilged ships these hydrodynamic terms are relatively small.

The free rolling motion is the solution of the roll equation above, i.e.

$$\phi = \phi_0 \cos \omega_n t$$

This indicates that the motion will continue forever with the same amplitude ϕ_0, but in reality we know this is not the case and the ship will eventually come to rest in the static equilibrium condition (in this case upright position), if not disturbed further by external or internal excitation, because the energy is dissipated through the action of 'damping' however small it may be.

Going back to the expression for T_n, we see that for small amplitude roll motion the natural frequency of the ship is a function of GM. Therefore, a vessel with high GM will have a jerky roll motion of a short period, while a vessel with low GM will show smoother roll motion of a longer period. Often smoother motion characteristics are preferred in practice, but it has to be paid for with reduced stability.

Using the relationship between T_n and GM, a ship in service may be quickly checked on its centre of gravity by simply observing the rolling motion of the vessel while stationary in a relatively calm water and consulting hydrostatics particulars for KM value. Indeed there are some commercial systems which measure the roll motions of offshore vessels in realistic sea conditions, analyse the spectral characteristics of the motions in real time to deduce the GM while the vessel is operating normally.

8.3 Basic Concept of Dynamical Stability

Now we shall consider the roll motion with external excitation moments present. The roll equation in this case is

$$I_\phi \ddot{\phi} = -\Delta g GZ(\phi) + M_E(\phi)$$

where $M_E(\phi)$ is the external heeling (or roll) moment as a function of the roll angle while its time-dependence is ignored.

Rewriting the equation,

$$I_\phi \ddot{\phi} + \Delta g GZ(\phi) = M_E(\phi)$$

Now

$$\ddot{\phi} = \frac{d^2\phi}{dt^2} = \frac{d\left(\frac{d\phi}{dt}\right)}{dt} = \frac{d\left(\frac{d\phi}{dt}\right)}{d\phi}\frac{d\phi}{dt} = \frac{1}{2}\frac{d\left(\frac{d\phi}{dt}\right)^2}{d\phi}$$

Substituting this into the motion equation and integrating both sides in ϕ from ϕ_0 to ϕ_1

$$\frac{1}{2}I_\phi \int_{\phi_0}^{\phi_1} \frac{d\left(\frac{d\phi}{dt}\right)^2}{d\phi} d\phi + \int_{\phi_0}^{\phi_1} g\Delta GZ(\phi)d\phi = \int_{\phi_0}^{\phi_1} M_E(\phi)d\phi$$

Since $\dfrac{d\left(\frac{d\phi}{dt}\right)^2}{d\phi}d\phi = d\left(\frac{d\phi}{dt}\right)^2 = d\dot{\phi}^2$,

$$\frac{1}{2}\left[I_\phi \dot{\phi}^2\right]_{\phi_0}^{\phi_1} + \int_{\phi_0}^{\phi_1} g\Delta GZ(\phi)d\phi = \int_{\phi_0}^{\phi_1} M_E(\phi)d\phi$$

Rearranging this equation, we get

$$\frac{1}{2}I_\phi \dot{\phi}_0^2 + \int_{\phi_0}^{\phi_1} M_E(\phi)d\phi = \int_{\phi_0}^{\phi_1} g\Delta GZ(\phi)d\phi + \frac{1}{2}I_\phi \dot{\phi}_1^2$$

where $\dot{\phi}_0$ is the roll velocity at angle ϕ_0, while $\dot{\phi}_1$ is the roll velocity at angle ϕ_1.

This is equivalent to saying that

initial kinetic energy	+	work done by external heeling moment	=	increase in potential energy	+	final kinetic energy

This is a simple statement of conservation of energy.

Let us now consider a specific case of $\phi_0 = \dot{\phi}_0 = 0$, i.e. the ship is kept steady at upright position, while being subjected to a steady external excitation moment, say wind on port beam. When the ship is released suddenly but without imparting velocity to it, i.e. $\dot{\phi}_0 = 0$, it will roll to leeward past the 'angle of equilibrium' ϕ_E, at which point the righting moment equals the heeling moment, and reach an 'angle of lurch' ϕ_L. There it turns back to windward direction past ϕ_E to an angle ϕ_2 where $0 < \phi_2 < \phi_E$. This process is repeated until the vessel rests at the static angle of heel ϕ_E.

Remember that we have assumed no damping, and in this case $\phi_2 = \phi_L$ and one can also see that $\dot{\phi}_0 = \dot{\phi}_L = 0$. The left hand side term of the integrated equation, therefore, is simply the area under the righting moment curve from $\phi = 0$ to ϕ_L. The right hand side term is also the area under the heeling moment curve within the same range of angles. Therefore, it can be summarised that

(area under the righting moment curve from $\phi = 0$ to ϕ_L) = (area under the heeling moment curve from $\phi = 0$ to ϕ_L).

This is shown graphically in Fig. 8.1.

The curve of area under the righting moment is called the **dynamical stability curve** and a typical example is shown in Fig. 8.2.

From the foregoing discussion we could define dynamical stability as 'the ability of a vessel to absorb the external heeling moment and convert it into potential energy'. This concept is significant in that it provides a means to interpret the statical stability curve in a more rational way.

It should be noted that the concept of dynamical stability is only concerned with the energy levels at initial and final conditions and it is not really interested in what goes on in between, i.e. the time is conveniently removed from the consideration. However, all dynamic systems are dynamic because they are time-dependent. Therefore we can say that calling this 'dynamical stability' is strictly speaking incorrect. For this reason it is sometimes called **quasi-dynamical stability**. Be that as it may, at least it takes the notion of stability into the realms of dynamical world. We shall examine some typical cases in Section 8.5.

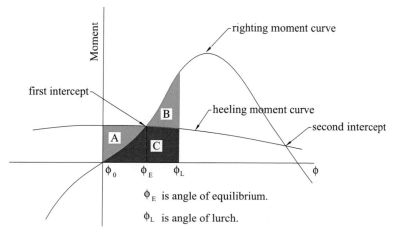

Fig. 8.1 Angles of equilibrium and lurch for $\phi_0 = \dot{\phi}_0 = 0$ (A + C = B + C)

Fig. 8.2 Dynamical stability
at an angle is the area under
the righting moment curve
up to that angle

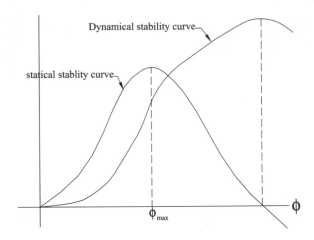

8.4   Heeling Moments

As has been mentioned a few times, righting moment in the stability context usually
refers to mass-moment. This can bring about some confusion, because the heeling
moment is normally calculated by multiplying the force by lever, resulting in force-
moment. In order to accommodate easy comparison between the righting moment and
heeling moment, therefore, it is necessary to convert these quantities into a common
format. The most convenient way of doing this is to use the equivalent levers. We
already know about the righting lever, but the heeling lever can be obtained by

(heeling lever) = (heeling moment)/$g\,\Delta$ if the heeling moment is force-moment.

The internal and external factors which create heeling moment include

wind
wave
passenger and cargo movement
ships turning at high speed
water on deck
mooring forces
current.

We shall examine heeling moments due to wind and cargo lifting over the side.

Wind

Since wind heeling moment is the major component of the so-called 'weather criteria'
which will be discussed in the next chapter, we shall examine it in some length.

The pressure due to wind of a velocity uniform in time and space on a flat plate
normal to the wind direction can be derived from Bernoulli's equation and is of the
form

$$P_w = \frac{1}{2}\rho_A U^2 C_D$$

where

ρ_A mass density of air (=1.25 kg/m^3)
U wind speed (m/s)
C_D drag coefficient.

However, wind speed does vary both in time and space. The time-dependence of wind speed can only be dealt with by introducing spectral analysis, but in the context of stability, it is often dealt simply with the so-called **gust factor** and we shall come back to it shortly. The space-dependence is quite complicated. For any realistic ship wind speed varies depending on the height above sea level of the ship's surface subjected to wind.

The wind speed is often cited for $z = 10$ m (where z is the height above the water level), and the relationship between the speed and height is often expressed in the form

$$\left(\frac{U_z}{U_{10}}\right) = \left(\frac{z}{10}\right)^{1/m}$$

where m is a constant and often 7 is used, U_z is the wind speed at z and U_{10} is the wind speed at 10 m above sea level.

For normal ships the critical wind direction is usually beam wind and we shall assume that it is always so.

Now, the total force on the vessel due to wind is

$$F_W = \int_A \frac{1}{2}\rho_A U_z^2 C_D dA$$

This will be a rather laborious process and so we can resort to discrete summing instead of integration.

The drag coefficient C_D has been introduced to account for the fact that windage surface of the same windage area may not necessarily experience the same wind force. This is due to differing shapes and orientation of the surfaces concerned and the value of C_D is often determined through experiments.

One more factor to consider and we shall have a fairly comprehensive picture, viz the matter of the wind force lever. Since we wish to compute wind heeling moment and we know only the force so far, the determination of the lever is the next step. Consider an arbitrary object whose shape, when projected on a plane normal to the wind direction, is as shown in Fig. 8.3.

Wind force on a thin horizontal strip of thickness dz_i at the height z above sea level is

$$dF_w = \frac{1}{2}\rho_A U_z^2 C_D l_i dz_i$$

Fig. 8.3 An arbitrary
windage surface

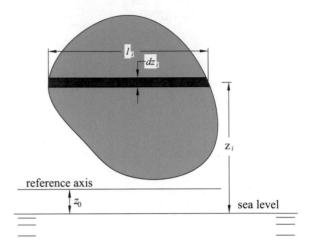

where l_i is the length of the ith strip.

The moment of this force about $z = z_0$ is

$$dM_w = dF_w(z_i - z_0) = \frac{1}{2}\rho_A U_z^2 C_D l_i (z_i - z_0)dz$$

Using the discrete summing,

$$M_w = \frac{1}{2}\rho_A C_D \sum_{i=1}^{n} U_i^2 l_i (z_i - z_0)dz_i$$

The heeling moment we thus calculated is for the ship at upright position. Consider the diagrams shown in Fig. 8.4.

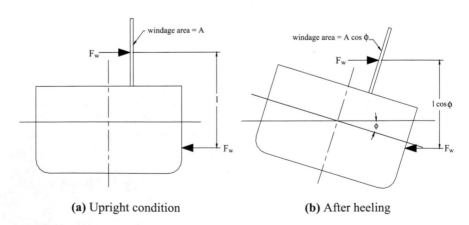

(a) Upright condition   **(b)** After heeling

Fig. 8.4 Wind heeling moment

As the ship heels over by an angle ϕ, the projected area becomes $A\cos\phi$ and the lever becomes $l \cdot \cos\phi$. Thus, the wind heeling moment as a function of ϕ can be expressed as

$$M(\phi) = M_0 \cos^2\phi$$

where M_0 is the wind heeling moment at upright condition.

As we discussed in the beginning of this section, we can then convert it into the **wind heeling lever** (or **wind heel lever**) as follows

$$l_w = M(\phi)\big/ g\Delta = M_0 \cos^2\phi\big/ g\Delta = l_0 \cos^2\phi\big/ g\Delta$$

where l_0 is the upright wind heel lever.

Note, however, that modern stability criteria tend to use a constant wind heel lever at all angles, which is erring on the safe side. Also for normal ships a constant wind speed is assumed for all heights and the time-dependence of wind speed is taken care of by using a gust factor, for which 1.5 is often taken. In other words gust heeling moment is taken to be 1.5 times that of steady wind. More of this in the next chapter.

Lifting of Heavy Weights over the Side

Again the heeling lever curve in this case can be superimposed on the statical stability lever curve. The heeling lever is often taken as

$$\frac{wa}{\Delta}\cos\phi$$

where

w mass of the object lifted
a transverse distance from centreline to end of boom
Δ displacement including w.

This formula is usually used by various authorities despite the fact that the lever may actually be larger than $(a\cos\phi)$ at certain heel angles.

8.5 Some Specific Cases

Steady Wind at Upright Condition

Imagine a ship being held upright by some means in the presence of steady beam wind. When the ship is released gently without imparting a velocity, the ship will start rolling to leeward increasing velocity all the time (potential energy being transformed into kinetic energy) until ϕ_E from which angle the velocity will start decreasing (kinetic energy being absorbed into potential energy) to the angle of lurch ϕ_L at which point all kinetic energy is absorbed into the potential energy (i.e. area A =

area B). If there is no damping, the windward rolling begins, going back to the upright condition at which point the whole process starts all over again. In reality, however, the two extreme angles will eventually converge to the angle of equilibrium ϕ_E. The intersections of the wind heel lever and righting lever curves are known as the first intercept and second intercept. This is illustrated in Fig. 8.5.

Initial Windward Heel

The ship is rolled over to windward reaching ϕ_0 due to, say, wave action whilst being subjected to a steady wind on the port beam. First of all, we ignore the effects of waves on the statical stability curve. The situation then is shown in Fig. 8.6.

In this case the vessel will heel to leeward up to ϕ_L so that area A is the same as area B, i.e.

$$\int_0^{\phi_L} [GZ(\phi) - l_w(\phi)]d\phi = 0$$

where

$l_w(\phi)$ is the wind heel lever and $l_w(\phi) = \frac{M_w(\phi)}{g\Delta}$
$M_w(\phi)$ is the wind heel moment.

However, if the area B up to the second intercept angle is smaller than A, the ship will capsize.

Initial Windward Heel Followed by Gust

In the case discussed above, if a gust blows just when the ship reaches ϕ_0, the situation is illustrated in Fig. 8.7. It is similar to the above, but the heeling moment will be replaced by the gust moment, which is usually taken as 1.5 times that of steady wind heeling moment, i.e.

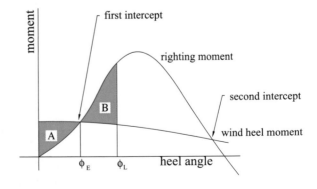

Fig. 8.5 A ship still in steady beam wind at upright condition

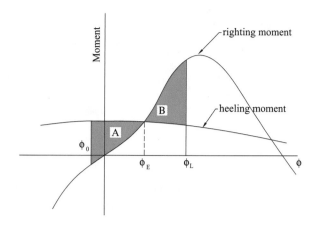

Fig. 8.6 The case of initial windward heel (area A = area B)

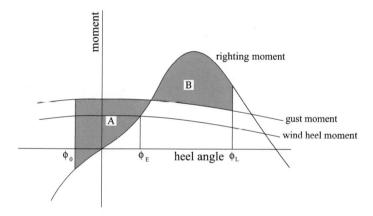

Fig. 8.7 Initial wind heel followed by gust

$$l_w(\phi) \int\limits_0^{\phi_L} \left[GZ(\phi) - l_g(\phi)\right]d\phi = 0$$

where $l_g(\phi) = 1.5l_w(\phi)$ and is the gust heel lever.

Negative Initial GM

Consider a vessel with a negative initial GM were released from its upright position in the absence of all heeling moment. If it is then released gently, the ship will initially heel over to ϕ_L so that area A is the same as area B, i.e.

Fig. 8.8 Angle of lurch for
negative initial GM

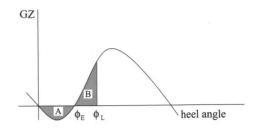

$$\int\limits_{0}^{\phi_L} GZ(\phi) = 0$$

The ship will then swing back and so on until finally it rests at the angle at which
the moment $=0$ (the angle of loll) as shown in Fig. 8.8.

Key Points

- Conservation of energy principle (between kinetic and potential energy)

$$\tfrac{1}{2}I_\phi \dot{\phi}_0^2 + \int\limits_{\phi_0}^{\phi_1} M_E(\phi)d\phi = \int\limits_{\phi_0}^{\phi_1} g\Delta GZ(\phi)d\phi + \tfrac{1}{2}I_\phi \dot{\phi}_1^2$$

- Gust heeling moment (lever) $=1.5$ time steady wind heeling moment
  (lever).
- Dynamical stability curve $= \int_0^\phi GZ(\phi)d\phi$

Exercises

1. A passenger vessel is operating at an even keel draught of 8 m. The righting lever
   curve of this vessel in its intact state can be expressed by

   $$GZ = 3.5\phi - 2.0\phi^3$$

   Due to steady wind on the port beam, the vessel heels at a steady angle of $10°$
   to leeward. At this position a gust blows from the same direction as the steady
   wind. Find the angle of lurch. State clearly any assumptions you make.

2. A cargo ship is operating in sea water at a displacement of 20,000 tonnes. In this
   operating condition the statical stability of the vessel can be expressed by

   $$GZ = 0.7\phi - 0.58\phi^3 + 0.1\phi^5$$

    (a) A cargo of mass 100 tonnes was moved from starboard to port by 10 m hor-izontally. Compute the new statical stability curve and indicate the approx-imate angle of loll.

    (b) As the ship is lying at the angle of loll found in (a), a gust blows from port beam. Estimate the angle of lurch. Assume the gust heel lever to be constant at 0.1 m at all angles of heel in this particular case.

3. A monohull vessel displaces 22,000 tonnes at the even keel operating draught of 9 m. At this draught the righting lever of the vessel is known to be expressed by

$$GZ = 3.5\phi - 2.0\phi^3$$

for the assumed KG of 10 m.

During the sea trial, the vessel made a turning manoeuvre at a linear speed of 6 m/s in clockwise direction along the circumference of a circle of 400 m radius. During this manoeuvre it was observed that the vessel had a steady heel of 6° to port. Assume that the ship experiences a centrifugal force at its centre of gravity and the resistance to it at half draught.

    (a) Estimate the actual KG.

    (b) In the steady condition stated above, the vessel rolled back to 2° starboard due to wave action. Find the consequent angle of lurch.

4. A prismatic vessel shown below has the following dimensions:

length $= 100$ m
breadth $= 20$ m
depth $= 15$ m
draught $= 7$ m even keel at sea
KG $= 5$ m

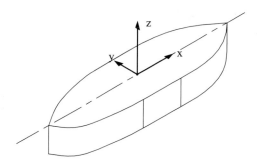

The waterline on the port side is expressed by

$$y = 10 \text{ for } -10 \le x \le 10$$

$$y = \frac{75}{8} + \frac{x}{8} - \frac{x^2}{160} \quad \text{for } 10 \le x \le 50$$

$$y = \frac{75}{8} - \frac{x}{8} - \frac{x^2}{160} \quad \text{for } -50 \le x \le -10$$

Derive an expression for the righting arm curve using the wall-sided formula up to the deck edge immersion angle.

5. The vessel shown in the above question has a gun mounted amidships, 3 m above the deck. If a projectile of mass 1 tonne was fired at a muzzle speed of 600 m/s when the ship was upright, what will be the angle of lurch of the vessel. The roll mass moment of inertia of the vessel is 450,000 tonne-m^2 and the resistance to the recoil can be assumed to be felt at half the draught (note: $\int \tan^2 \phi \sin \phi d\phi = \sec \phi + \cos \phi + C$).

6. An ocean-going ship of displacement 25,000 tonnes at the draught of 8 m is being designed. The righting lever curve for this condition can be expressed as

$$GZ = 0.35\phi - 0.32\phi^3$$

where ϕ is the angle of heel in radians.
The above-water profile of the ship is given in the figure below.
The ship is subjected to a steady wind pressure of 0.0514 tonnes/m^2 acting at right angles to the ship's centreline.

(a) Estimate the steady wind heel lever and the gust heel lever.
(b) From the resultant angle of equilibrium in the steady wind, the ship rolled **by** 12° to windward due to wave action. The ship is then subjected to a gust. Estimate the resultant angle of lurch.

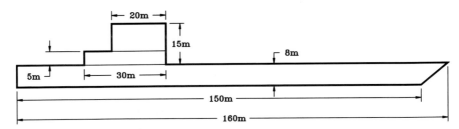

Above-water profile of the ship
(Not to scale)

Chapter 9
Intact Stability Criteria

9.1 Introduction

Over the centuries many ships have been lost at sea through capsizing either because of operational errors and/or lack of inherent stability of ships whether intact or damaged. In the past, even though people were concerned about the maritime safety, particularly those whose lives and livelihood depended on it, certain degree of risk was accepted as inherent in maritime activities. Venturing out to sea on a ship was regarded as an 'adventure'. In risk-averse modern times, however, such cavalier attitude towards ship safety is no longer acceptable and people have endeavoured to set acceptable standards.

Fairly comprehensive understanding of intact ship stability was achieved by mid-19th Century including dynamical stability concept, much of which we have studied in the foregoing chapters. However, judging whether a ship has sufficient stability or not was left largely to the basic concept of initial GM and even some naval architects advocated leaving the task to ships' masters, citing too many operational conditions to produce clear-cut criteria as an excuse. It is unnecessary to repeat the fact that ships must have sufficient stability to prevent capsizing. The main difficulty here is to judge what is 'sufficient'. Despite the understanding of the nature of stability and process of capsizing, it is much more difficult to establish a standard to enable sound judgement. This was one of the reasons why the first internationally accepted stability regulations did not appear until mid-20th Century.

Intact stability regulations thus produced have always been prescriptive until fairly recently. In other words, the regulations specified one or a number of criteria and demanded ships of certain categories satisfy them. This necessarily implied that the authorities regarded ships satisfying these criteria were 'safe' as far as their intact stability was concerned. However, deciding on these criteria is not easy, as lack of knowledge in the phenomenon of capsize has to be compensated for by making them more strict, but over-restrictive regulations will put shipping at an unnecessary disadvantage.

© Springer Nature Singapore Pte Ltd. 2019
B. S. Lee, *Hydrostatics and Stability of Marine Vehicles*, Springer Series on Naval Architecture, Marine Engineering, Shipbuilding and Shipping 7,
https://doi.org/10.1007/978-981-13-2682-0_9

At first, the criteria were determined by looking at the statistical data of past ships. As the knowledge increased in this area, however, criteria became more rational, albeit painfully slowly due to political and commercial reasons. In most countries only the public outcries after major high-profile accidents forced the authorities to bring about the changes. It can be said that the modern intact stability criteria which are more logical and fit for purpose at last are the painful result of past accidents and many lives lost at sea.

We shall examine some of these criteria.

9.2   Criteria Based on Statistics

In 1939, in his now famous Ph.D. thesis, Rahola produced a set of intact stability criteria which has a logical basis for the first time. Until this what criteria there were only specified minimum initial GM, and, therefore, the 'Rahola Criteria' marked an important point in ship safety. His work was based on the information of 34 vessels known to have capsized. The stability characteristics of these vessels were analysed and by comparing them with those which did not capsize, he produced the statical stability criteria as shown in Fig. 9.1. Note that these criteria include certain requirements on the righting lever curve as well as initial GM value.

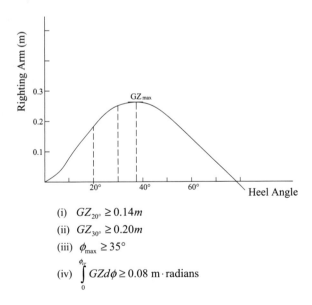

(i) $GZ_{20°} \geq 0.14m$

(ii) $GZ_{30°} \geq 0.20m$

(iii) $\phi_{max} \geq 35°$

(iv) $\int_{0}^{\phi_{cr}} GZd\phi \geq 0.08 \text{ m} \cdot \text{radians}$

Fig. 9.1   Rahola criteria

The critical angle, ϕ_{cr}, refers to the smallest of ϕ_m, ϕ_f, 40° and the angle of heel at which cargo shifting can occur. ϕ_m is the angle at which GZ is maximum, and ϕ_f is the downflooding angle, or the angle at which any opening in the hull, superstructures or deckhouses that cannot be closed weathertight is immersed leading to progressive flooding. Small openings through which progressive flooding cannot take place need not be considered as open.

People took note of this development, but the second world war intervened. However, even when the war was over, the international community was very reluctant to adopt such criteria for a variety of reasons with many nations preoccupied in rebuilding their economy. After much pressure from the leading industrial maritime nations, the then IMCO (Intergovernmental Maritime Consultative Organisation—now IMO) moved to set up a sub-committee on subdivision and stability in 1962. The Committee followed essentially the same approach as Rahola's a quarter of a century earlier, and collected information on 47 vessels known to have capsized from various countries. These were then analysed using the following set of parameters:

GM_0 initial GM
ϕ_m angle of maximum GZ
ϕ_v vanishing angle
GZ_m maximum GZ
GZ_{20} GZ at 20° heel
ϕ_f down flooding angle
ϕ_d angle of deck edge immersion.

An example of the resulting criteria is shown in Fig. 9.2. They have been claimed to be unreliable and insufficient, particularly in the lower size range close to the 24 m lower length limitation and for vessels operating in light condition. Some vessels satisfying IMCO criteria, sometimes with ample margins, capsized, which would appear to support these criticisms.

9.3 Criteria Based on Dynamical Stability (Weather Criteria)

In 1962 the US Navy adopted a 'weather criterion' (sometimes called 'severe weather criterion' or 'wind line criterion'). This was based on the concept of dynamical stability and an initial windward heeling of 25° from the angle of static equilibrium was assumed (see Fig. 9.3). The situation with an initial windward heeling has already been examined, and further explanation is unnecessary here, but note that the criterion demands dynamical stability margin of 40%. The figures of 25° and 40% are arbitrary and these, particularly the former, attracted some criticisms. Nevertheless, the track record shows that US naval vessels satisfying this criterion appear to have adequate stability. After all, the proof of stability criteria lies in its track record, although it must be remembered that overly stringent requirement can restrict the design flexibility and hence the operational advantage of the subject ships.

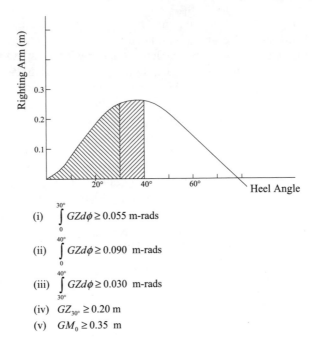

(i) $\displaystyle\int_{0}^{30°} GZd\phi \geq 0.055$ m-rads

(ii) $\displaystyle\int_{0}^{40°} GZd\phi \geq 0.090$ m-rads

(iii) $\displaystyle\int_{30°}^{40°} GZd\phi \geq 0.030$ m-rads

(iv) $GZ_{30°} \geq 0.20$ m

(v) $GM_0 \geq 0.35$ m

Fig. 9.2   1968 IMCO criteria for fishing vessels

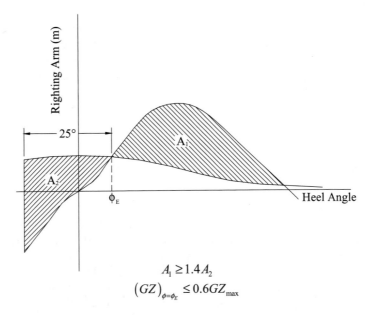

$$A_1 \geq 1.4 A_2$$
$$(GZ)_{\phi=\phi_E} \leq 0.6 GZ_{max}$$

Fig. 9.3   1962 US Navy weather criterion

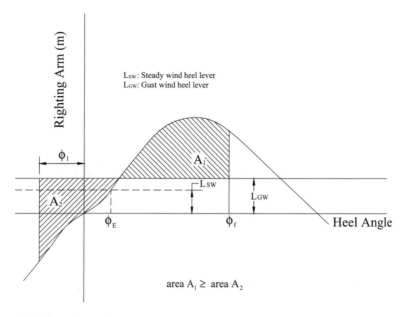

Fig. 9.4 USSR weather criterion

The weather criterion adopted by the USSR, shown in Fig. 9.4, is similar to the US Navy criterion, but there are some differences.

(a) This criterion introduces gust heeling moment. The time-dependence of wind is conveniently, albeit simplistically, dealt with by introducing a gust factor of 1.5. It is also assumed that the gust heeling moment is purely potential.
(b) No reserve dynamical stability is required by the criterion.
(c) The wind heeling moment is assumed to be a constant rather than a function of heel angle.

When IMCO finally arrived at the statical stability criteria in 1968, it did not reject the other approach, i.e. weather criterion. In 1983 the IMO Maritime Safety Committee made a recommendation of a 'severe wind and rolling criterion' for the intact stability of passenger and cargo ships over 24 m in length. This was ratified by the General Assembly of the IMO in 1985 to be the recommended criterion. Since this is the current criterion, it is worth examining this.

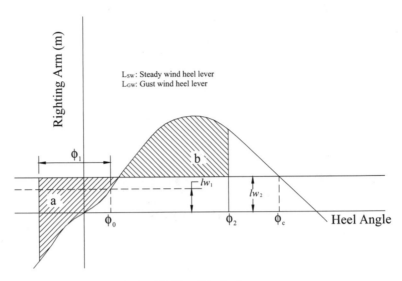

Fig. 9.5 IMO 1985 'Severe Wind and Rolling Criterion'

The essential features of the criterion is illustrated in Fig. 9.5, but the verbatim (more or less) quotes of the key part of the criterion is as follows:

The ability of a ship to withstand the combined effects of a beam wind and rolling should be demonstrated for each normal condition of loading, with reference to Fig. 9.5, as follows:

(a) The ship is subjected to a steady wind pressure acting perpendicular to the ship's centreline which results in a steady wind heeling lever (lw_1).
(b) From the resultant angle of equilibrium ϕ_0, the ship is assumed to roll due to wave action to angle of roll ϕ_1 to windward.
(c) The ship is then subjected to a gust wind pressure which results in a gust heeling lever lw_2.
(d) Under these circumstances, area b should be equal to or greater than area a.

The angles in Fig. 9.5 are defined as follows:

ϕ_0 = angle of heel under action of steady wind
ϕ_1 = angle of roll to windward due to wave action
ϕ_2 = angle of downflooding (ϕ_f) or 50° or ϕ_c whichever is less.
ϕ_c = angle of second intercept.

The wind heeling levers are constant values at all angles of inclination and should be calculated as follows:

$$lw_1 = \frac{P \cdot A \cdot Z}{\Delta} \text{ (m)} \quad \text{and} \quad lw_2 = 1.5\,lw_1 \text{ (m)}$$

where

$P = 0.0514$ t/m^2 (t$= 1000$ kg)
A $=$ Projected lateral area of the portion of the ship and deck cargo above the waterline (m^2)
Z $=$ Vertical distance from the centre of A to the centre of the underwater lateral area or approximately to a point at one half the draught (m).
$\Delta =$ Displacement (t).

Then a fairly complex procedure of determining the windward heel angle, ϕ_1, follows. A number of tables are provided, so that the rolling characteristics of the vessel concerned can be taken into consideration. For example, ships with round bilges, those with sharp bilges, and those with bilge keels are differentiated. Some form factors such as B/T ratio and the block coefficient are also taken into account, but, strangely, no favour is given to any active anti-rolling device.

These weather criteria often differ from each other in details, but they are all based on more or less severe scenarios of ship motion and wave/wind actions.

9.4   Comparison Between Statistical and Dynamical Approach

The statistical criteria are in general simple to use and the calculation involved in applying the rules is minimal. They are based on operational experience and some data can be obtained from rigorously conducted model experiments, and therefore can be argued to be realistic. However, they are subjected to a number of criticisms. Firstly, the data used to compile the statistics vital in deriving the criteria may be considered heterogeneous as to the type, age and size of the vessels that capsized as well as the operational conditions when the accidents occurred. Unfortunately, or rather fortunately, the capsize incidents are quite few and far between, making the statistical samples rather small. There is also the danger that some of these incidents were reported wrongly or not at all, making the data somewhat unreliable and inaccurate. One can also argue that some ships which may capsize in the future have not done so until the time the sample was taken. All these can make the resulting criteria unreliable. Furthermore, one of the crucial shortcomings of this type of criteria is the fact that they cannot be readily applied to any novel design and ship type.

Severe weather criteria, on the other hand, are based on physics, although relying on a hugely simplified dynamic model. Therefore, it can be said that they are more logical and applicable to completely new designs and any operating conditions without much adjustment. Despite the fact that most dynamical theories tend to be quite complex, the weather criteria are in general easy to apply and understand. A really attractive feature of these criteria is the fact that they are flexible for further refinement incorporating future theoretical advancement, although the current IMO severe weather criterion does not take into account the effects of waves and other detrimental factors such as water-on-deck. Often the main parameters are chosen rather arbitrarily, but it has much better track record compared to the statistically derived criteria.

Despite some shortcomings of the weather criteria, therefore, one could say that the intact stability is now reasonably secure. The pursuance of further safety lies in stability of ships in damaged conditions and minimizing operational errors and human mistakes, and the damage stability is the subject of the next chapter.

9.5   Trim and Stability Particluars

It is a statutory requirement for all merchant vessels and many leisure craft to be to have a trim and stability booklet on board. It contains all information necessary for the ship's master and deputies to ensure the vessel has sufficient intact stability in all possible loading conditions. The regulations make certain information mandatory, and samples are sometimes provided. Smaller craft will have relatively short 'booklets' while the documents provided on board merchant ships tend to be bulky volumes. Whatever the size, such documents contain all hydrostatic, stability and bending moment data for all loading conditions that the ship is expected to encounter, as well as the detailed instructions on how to use the information and what the relevant regulations are. An example table of contents of a merchant ship is given in Table 9.1.

It is worth noting that there are now many commercial software available which can be tailor-made to any particular ship and run real time on onboard computers. Most merchant ships have them in addition to the trim and stability booklet. Although it is by no means mandatory, IMO does have a guidance on the provision and use of this type of software. The ship's officers can use this software to predict the ship's attitude and stability for a given loading condition as well as the structural loading the ship will experience.

Table 9.1 Table of contents of a trim and stability booklet of a bulk carrier

1.1	OWNERS PREAMBLE / GENERAL INTRODUCTION
1.2	INSTRUCTION TO THE MASTER
2	MAIN PARTICULARS
3	DEFINITIONS AND CONVERSION TABLE
4	NOTES REGARDING STABILITY AND LOADING OF THE SHIP
4.1	Stability
4.2	Curves of righting levers
4.3	Free surface effects
4.4	Hydrostatic and isocline stability data
4.5	Use of tables
4.6	Trim and draught restrictions
4.7	Immersion of Propeller
4.8	Air draught
4.9	Visibility
4.10	Draught and trim calculation
4.11	Displacement from draught readings
4.12	Minimum angle of flooding
4.13	Intact stability criteria
4.14	Effect of wind and rolling
4.15	Demands regarding damage stability
4.16	Ballast Water Exchange
4.17	Load and strength calculation
4.17.1	Limitations due to strength in flooded conditions
4.17.2	Allowable still water shear force
4.17.3	Correction to actual shear force
4.17.4	Allowable still water bending moments
4.17.5	Sloshing
4.17.6	Loading of timber deck cargo
4.17.7	Light ship weight and distribution
5	WORKING EXAMPLE
6	TANK AND CAPACITY INFORMATION
7	LOADING CONDITIONS
8	HYDROSTATIC DATA AND PLOT OF THE HULL
9	STABILITY DATA (MS AND KN TABLES)
10	MAXIMUM KG AND MINIMUM GM LIMIT CURVES
11	RELEVANT OPENINGS AND POSITIONS / FLOODING ANGLES
12	LIGHTWEIGHT DISTRIBUTION
13	INCLINING EXPERIMENT
14	EMPTY CALCULATION SHEET
15	DRAWINGS
16	BC-A LOADING CONDITIONS

Chapter 10
Basic Concept of Damage Stability and Watertight Subdivision

10.1 Introduction

Damage to a vessel which compromises the watertight integrity of the hull will lead to ingress of water into the compartment(s) of the vessel. The flooding consequent to the broaching of the watertight skin of the ship will affect the attitude of the vessel, i.e. trim, draught and heel, and the stability characteristics will be affected, usually for the worse. The 'remaining' stability after sustaining damage is known as *residual* or *damage* stability.

All types of vessels are subject to risk of being lost if they are damaged whether by collision, grounding or internal mishaps such as fire and explosion. Such accidents are frequent enough in practice that some degree of protection against the eventualities of flooding should be given. For example, sufficient residual stability should be provided so as to maximise the chance of survival of the passengers/crew and ultimately the vessel and cargo. One effective way of achieving this is dividing the internal space of the ship into a number of watertight compartments; a practice known as *watertight subdivision*.

One of the difficulties in doing this is the fact that damages are not planned (as distinct from the design activities which work on planned state of affairs), and thus are unpredictable. It is not known where the damage will occur and what the extent will be—indeed it is unknown whether the vessel will sustain any damage during its lifetime at all. This means that we have to consider the effects of possible damage scenarios and their probability of occurrence. The probability of occurrence has been traditionally incorporated into the damage stability regulations as a multiplication factor and such regulations are said to have adopted the *factorial system* in the practice of watertight subdivision. However, with the industry having taken a step towards *goal-setting* rather than *prescriptive* regulatory regime, the so-called probabilistic assessment of survivability of ships and passengers/crew is in vogue at present time.

© Springer Nature Singapore Pte Ltd. 2019
B. S. Lee, *Hydrostatics and Stability of Marine Vehicles*, Springer Series on Naval Architecture, Marine Engineering, Shipbuilding and Shipping 7,
https://doi.org/10.1007/978-981-13-2682-0_10

Whatever the stability regulatory regime, we need to have some methods of dealing with various damaged situations. Clearly, however, we need to understand first in what way the ship is affected after a damage. We can consider here only the outwardly visible effects which can be readily measured directly as follows:

Draught

The draught will change so that the displacement of the remaining unflooded part of the ship is equal to the displacement of the ship before damage less the weight of any liquids or other cargo which are lost to sea.

Trim

The underwater body shape changes and the flooded water also causes the centre of gravity to change. It is obvious that a new state of equilibrium will involve an alteration in trim.

Heel

If the flooding is asymmetrical, a static heel will be inevitable. A static heel poses a danger of capsizal and, therefore, regulations demand that in such cases cross-flooding arrangement be made so that port and starboard balance can be achieved within a short time. Nevertheless, it is important to note that until such a balance can be achieved, the ship may go through interim state of static heel.

Stability

It is reasonably obvious why the stability characteristics will alter, and this will be discussed in more detail later.

Freeboard

One of the unpleasant consequence of the increase in draught is the reduction in the amount of freeboard available. Freeboard is a kind of reserve buoyancy and stability, and its reduction can be serious. For example, the so-called *margin line* is used to define the minimum freeboard allowed in any damaged state.

Loss of Ship

Where changes to any one or more of the above factors are excessive beyond certain limits, loss of ship can occur, either through capsizal, plunging, sinking, structural breaking up or any combination of these processes.

It can be seen that the hydrostatics of a damaged ship is somewhat more complicated. Essentially, however, it is still a floating body, and, as such, exactly the same principles as used in intact hydrostatics can be applied. The subject of damaged stability and the safety or survivability of damaged ships is a complex one and can easily form a separate book. The discussion in this chapter, therefore, is necessarily brief and only deals with the most basic ideas.

10.2   Intermediate Stages of Flooding

During the intermediate stages of flooding (i.e. from the time of damage to the time when a final new equilibrium has been reached) water is continually flooding into the damaged compartment(s). Flooding may not be 'continuous', as the water flow is determined by the difference in the head of the water surface inside and out. On the outside the surface is continuously changing due to wave action, while inside the water already flooded in will *slosh* about due to the vessel motion. Therefore, the water flow may not even be one-way all of the time.

Nevertheless, short of tackling this with a time-domain simulation, some simplification is necessary and usually the following assumptions are made:

– The vessel is assumed to be in static equilibrium at every stage of flooding, with water surface in the flooded compartment parallel to the surface of the sea but at a lower level.
– In the case of asymmetrical flooding involving spaces cross-connected by pipes, ducts, etc., it is normally assumed that the flooding water in the damaged wing spaces reaches sea-level.
– During the intermediate stages of flooding, heel may occur either as a result of negative residual GM or from asymmetrical flooding. Since we are dealing with a transient state, some heel is acceptable provided it is not excessive and long-lasting and also the range of stability and maximum righting arm include sufficient margin of safety.

10.3   Equilibrium, Draught and Stability After Flooding

In order to calculate the ship's attitude in equilibrium, broadly speaking two methods are used representing perhaps the two fundamental ways of looking at the flooded water: lost buoyancy method (LBM) and added weight method (AWM).

Lost Buoyancy Method
In this approach the damaged compartments are treated to be completely open to sea, and, therefore, they no longer contribute to the ship's buoyancy. Just imagine the normally watertight hull being suddenly changed to a kind of sieves with large holes! The compartments flooded are consequently removed completely from the hydrostatic calculations. To compensate for this lost buoyancy, the ship has to sink further to an increased draught until the displacement becomes identical to that prior to the damage, as the ship's mass is assumed to be unchanged. The centre of gravity of the ship remains unchanged.

It is easy to see that this is a physical over-simplification, but the method is simpler to use than the added weight approach, and consequently most regulations rely on this method.

Added Weight Method

In this approach the damaged compartment is treated to be still intact, and the flooded water is treated in exactly the same manner as ballast water taken on board, except that the amount of flooded water changes with the draught. In other words the mass displacement of the ship has been increased by the amount of water ingress. An increase in draught occurs to provide the extra buoyancy for this added weight of flood water. This, of course, will alter the centre of gravity of the vessel as well.

The method is a little more complex than the LBM, but some people consider it represents the reality a little more closely. However, the actual physical phenomenon of flooding due to damage is somewhere between these two extreme models. It is useful for dealing with the intermediate stages of flooding and also when we wish to work out the length and location of the compartment which can be flooded without the waterline going over the margin line (*floodable length* calculation).

It is important to note that both methods should produce identical values of the physically measurable quantities. For example, draught and righting moment are some of the items which can be measured, while the location of centre of gravity cannot be measured as it is essentially a conceptual point. The main points of the difference between the two approaches are summarised below:

	LBM	AWM
Displacement	Δ	$\Delta + \delta\Delta$
VCG	$(KG)_0$	$(KG)_1$
Draught	$T + \delta T$	
VCB	$(KB)_1$	$(KB)_2$
Righting moment	R.M.	
Righting arm	$R.M./\Delta$	$R.M./(\Delta + \delta\Delta)$
Free surface effects	n.a.	Yes

It is crucial, therefore, to understand that residual GM value alone is insufficient information unless accompanied by a displacement used (or more commonly the method used to calculate it, i.e. LBM or AWM).

Before we examine the two methods using some examples, we need to determine how much water a damaged compartment can admit into it. This point is dealt with by a factor called *permeability*, which is defined as the ratio of the volume which can be occupied by flood water to the total nominal volume of the compartment and is usually denoted with a symbol μ. Typical values of permeability are:

accommodation spaces 95%
machinery compartments 85%
coal bunkers, stores, cargo holds 60%
tanks 0–95%.

Of course, cases can be made for differing values of permeability depending on circumstances.

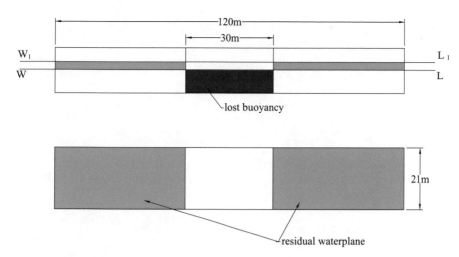

Fig. 10.1 Lost buoyancy approach

Example 10.1

To illustrate the general effects and principles, we consider a box-shaped vessel as being the only one in which the numerical work is not unduly laborious. Take a box-shaped vessel, 120 mL × 21 mB × 8 m even keel draught with KG = 6.0 m. A compartment of length 30 m and spanning the whole breadth of the vessel, situated amidships is flooded due to a damage to the hull. Assume 100% permeability. Calculate the residual GM.

Solution

Prior to the damage occurring

$$\nabla = 120 \times 21 \times 8 = 20.160\,\text{m}^3$$

$\Delta = 20.160 \times 1.025 = 20{,}664$ tonnes
original WPA $= 120 \times 21 = 2520\,\text{m}^2$
We will do this with both methods, but with LBM in the first instance.

LBM

The idea is illustrated in Fig. 10.1.
  lost WPA $= 30 \times 21 = 630\,\text{m}^2$
  therefore, intact part of WPA $= 2520 - 630 = 1890\,\text{m}^2$
  residual TPC $= 1890 \times 1.025/100 = 19.37$ tonnes/cm
  lost buoyancy $= 30 \times 21 \times 8 \times 1.025 = 5166$ tonnes
  sinkage $= 5166/1937 = 2.667$ m
  new draught $= 10.667$ m and since the ship is box-like,
  new KB $= 5.334$ m
  residual waterplane $I_T = \frac{(120-30) \times 21^3}{12} = 69457.5\,\text{m}^4$

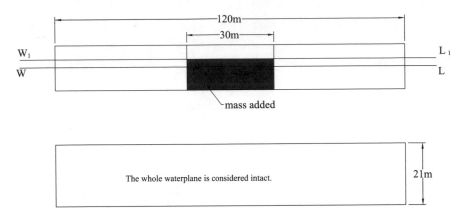

Fig. 10.2 Added weight approach

new BM $= \frac{69457.5}{20,160} = 3.445$ m
new KM $= 5.334 + 3.445 = 8.779$ m
KG does not change and is 6.00 m
Therefore, new GM $= 8.779 - 6.00 = 2.779$ m
Remember, the displacement is still 20,664 tonnes.

AWM

Unlike in LBM where the lost buoyancy is known, the added weight in AWM is indeterminate to begin with, because the amount of flooded water depends on the final draught. Usually the new draught is found through an iterative method, but in this particular case this problem is much simpler because the ship is box-like. Let the new draught be T. Remember the hull is considered to be still intact and the flooded water is treated as a weight added to the ship. The situation is illustrated in Fig. 10.2.

Therefore, the original displacement plus the added weight is the new displacement,

$120 \times 21 \times T = (120 \times 21 \times 8) + (30 \times 21 \times T)$

Solving it for T we get

T $= 10.667$ m and consequently KB $= 5.334$ m

Now we must obtain the new KG. The mass of the flooded water (up to T $= 10.667$ m) is

$\delta\Delta = 30 \times 21 \times 10.667 \times 1.025 = 6888.2$ tonnes
New $\Delta = 20,664 + 6888.2 = 27552.2$ tonnes
New $\nabla = \frac{27552.2}{1.025} = 26880.2$ m^3
new $KG = \frac{20,664 \times 6 + 6888.2 \times 5.334}{27552.2} = 5.833$ m
I_T of waterplane $= \frac{120 \times 21^3}{12} = 92,610$ m^4

$$BM = \frac{92,610}{26880.2} = 3.445 \, \text{m}$$

The transverse second moment of the free surface of the flood water is

$$i_T = \frac{30 \times 21^3}{12} = 23152.5 \, \text{m}^4$$

Therefore, free surface effect $= \frac{23152.5 \times \rho}{27552.2} = 0.861 \, \text{m}$
Finally,
residuary G $= 5.334 + 3.445 - 5.833 - 0.861 = 2.085$ m
As can be seen from this example the residual GM values calculated by the two methods are not the same. However, the righting moment which is directly proportional to GM at small angles of inclination should be the same. Therefore we should get identical values of (displacement $\times$ GM).
From the LBM 20,664 $\times$ 2.779 = 57,425 tonnes/m
From the AWM 27552.2 $\times$ 2.085 = 57,446 tonnes/m.
The small discrepancy is due to accumulated calculation errors.

When the flooding is not symmetrical, there will be changes in trim and heel as well as GM. This makes the added weight method quite clumsy to use, as this method requires the flood water to be calculated but it is variable depending upon the final attitude of the ship, which is dependent on the amount of flood water and so on. With the LBM, it is quite easy to calculate the lost buoyancy, and thus the sinkage to a new draught. Since the centre of buoyancy will have moved in both longitudinal and transverse directions, changes in heel and trim will occur. This can be calculated treating the ship as an arbitrary three-dimensional object floating in water. Note that the centre of flotation will no longer be at the centreline of the ship.

10.4 Factorial System of Subdivision

One of the very effective ways of improving the chance of survival for a damaged ship is to contain the flood water within certain areas in the vicinity of the water ingress. This can be achieved by dividing the internal space of the ship into a number of watertight compartments. Although the internal space is naturally divisible according to the uses, such as accommodation spaces, engine rooms, cargo spaces and so on, the main purpose of subdivision is to enhance the survivability of the ship after flooding. Therefore, a ship cannot have excessively long watertight compartments, however much it is desirable to have such spaces.

The traditional way of determining the length of watertight compartments is by determining the maximum length of the compartments allowed for each point along the longitudinal position of the ship. This is known as *floodable length* and is deter-

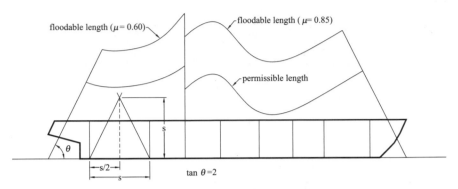

Fig. 10.3 An example of floodable and permissible length curves

mined from the requirement that the final equilibrium waterline must not immerse the *margin line* as well as maintaining the minimum required damage stability standard. An illustration of floodable length curve is given in Fig. 10.3. Note that the floodable length curve can be discontinuous at predetermined bulkheads which separate two compartments with differing permeability, such as the engine room and a cargo hold.

The margin line defines the highest line that the trimmed and heeled waterline is allowed to reach. It is determined in general 76 mm below the top of transverse watertight bulkhead, but can be modified by the presence of superstructures.

The relevant regulations also define the subdivision factor F. It is a function of the length of the ship used for subdivision and service that the ship is engaged in. Longer ships have lower F and ships which are more like passenger ships also have lower F. It is always less than 1.0.

The maximum subdivision length allowed (*permissible length*) is then obtained by multiplying the floodable length with the subdivision factor,

P.L. = F.L. × F

An illustration of permissible length is shown in Fig. 10.3 with the subdivision factor of about 0.6 for an example. Both F.L. and P.L. are usually plotted on the profile of the ship with the same scale on the horizontal and vertical axes. In this way, one can easily determine the maximum allowable extent of the compartment with its centre point at a given longitudinal location can be found by drawing a line from the permissible length point of the location at an angle of $\tan^{-1} 2$. Conversely, a proposed compartment can be tested by drawing similar lines from the locations of the two end bulkheads. If the intersection of these lines lies below the P.L. curve, the compartment meets the requirement.

In 1929 the International Conference on Safety of Life at Sea (SOLAS) adopted the factorial system of subdivision as outlined above. However, the regulations involving floodable length were removed from SOLAS in 2006 and replaced by the requirements to satisfy the new damage stability regulations. The basic idea of these regulations will be briefly examined in the next section.

10.5 Damage Stability Regulations

Capsize and sinking of ships is of major concern to the authorities trying to ensure the safety of human lives and ships, but it has other implications, such as marine pollution. It is hardly surprising, therefore, that damage stability of ships is regulated by International Bulk Chemical Code (IBC), International Gas Carrier Code (IGC), International Convention for the Prevention of Pollution from Ships (MARPOL) and International Convention on Load Lines (ICLL) as well as SOLAS, all under the aegis of International Maritime Organisation (IMO). They have different focus in regulating damage stability, and, therefore, the requirements and applicability are all slightly different. For example, ICLL provisions are applied to oil and chemical tankers and bulk carriers, while MARPOL regulations are primarily interested in oil tankers. SOLAS regulations are applied to some bulk and container carriers but primarily to passenger-carrying ships, such as ferries and passenger ships.

It is also interesting to note that the damage stability regulations in IBC, IGC, MARPOL and ICLL are deterministic, while only SOLAS adopts the probabilistic regulations. The basic concept of these two approaches will be briefly outlined below.

(a) **Deterministic Damage Stability Regulations**

These regulations specify the residuary stability requirements for damage virtually anywhere on the ship with predefined longitudinal, transverse and vertical extent of damage. Note that a damage occurring at the position of a bulkhead will normally result in two compartments flooding, except engine rooms which are only considered for single compartment flooding. The requirements include static equilibrium heel angle, maximum residuary GZ, positive range of stability from the equilibrium angle and the area under the residuary GZ curve. For example, MARPOL requires the following:

1. The final waterline, taking into account sinkage, heel and trim, must be below the lower edge of any opening through which progressive flooding (downflooding) may occur.
2. Static equilibrium heel angle for asymmetric flooding must be 25° (maybe extended to 30° if no deck edge immersion occurs) or less.
3. Positive residual righting lever up to minimum 20° from the equilibrium angle with the maximum righting lever of at least 1 m.
4. The area under the righting lever curve within the range above must be not less than 0.0175 m-radians.

(b) **Probabilistic Damage Stability Regulations**

The idea of assessing damage stability based on the probability of survival after suffering damage was proposed as early as 1950s. Due to lack of confidence in such a concept and other reasons, such as political, about 40 years had to elapse before the international community tentatively started using the so called probabilistic damage stability criteria. It was first inserted in the 1992 amendment of SOLAS 74, but its full-blown implementation had to wait until the adoption of a new amendment in 2005 when the Maritime Safety Committee of IMO resolved to adopt harmonized damage stability regulations which came into force in 2009 and consequently is now known as SOLAS 09. It is applicable to both dry cargo ships and passenger-carrying ships. It marked a significant departure from the old deterministic rules in that its spirit is more goal-setting than prescriptive.

The main idea is simple enough. A set of damage scenarios (damage cases) are identified each with its own probability of occurrence. For each damage case the probability of survival of the ship after sustaining the damage is calculated. The survivability is then multiplied to the probability of occurrence for each damage case and then summed for all cases. This produces the *attained subdivision index* (A) for the ship and the rules require this to be higher than the *required subdivision index* (R). R is determined from the length of the ship used for subdivision and the number of passengers carried.

In reality it is a complex process, primarily because the sum of the probabilities of all damage cases must equal 1, i.e. all possible damage cases must be considered. Furthermore, each damage location must be examined for the probability of the next bulkhead (be it transverse or horizontal) not to be breached. The computation is complicated and arduous. The rules are so complex that they need to be read with extreme care and attention to details. Indeed special training programmes had to be run to familiarise the designers with the new rules. A new service of calculating the attained subdivision index has also become a part of ship design consultancy companies.

Despite some of criticisms, the probabilistic rules are considered more rational than the deterministic ones which used to judge a ship either to pass or fail the criteria. The designers may also aspire to achieve a higher A than required by R to enhance survivability. Apart from all this, they do give designers a high degree of flexibility in placing bulkheads and creating compartments which is particularly useful for some passenger ships such as cruise ships.

With the increasing operational experience, the details of the probabilistic damage stability rules will evolve constantly moving towards safer ships, but it is certain that they are going to be with us for the foreseeable future in their current basic form.

Numerical Answers to the Exercises

Chapter 2

1. Draught = 6.631 m, pressure at the bottom = 60.62 kN/m^2
2. Volume to be increased = $\frac{\Delta}{\rho_f} - \frac{\Delta}{\rho_s}$ (m^3)

   $A_w = 100 \times TPC_{sw}/\rho_s$ (m^2)

   $d = \frac{\text{volume to be increased} A_w}{A_w}/100$ (cm)
4. $C_m = 0.984$, $C_p = 0.707$
5. Disp = 15,507.58 m^3

   $A_m = 160.574$ m^2

   $C_p = 0.754$
6. L = 88.235 m

   B = 11.268 m

   T = 3.192 m
7. $C_b = 0.56$

   $C_p = 0.609$

   TPC = 20.5 tonnes/cm

   New T = 4.878 m
8. $\nabla = 15605.58$ m^3

   $\Delta = 15,995.72$ tonnes

   $C_p = 0.707$

   In fresh water

   T = 8.210 m

   $C_b = 0.652$

   $A_m = 181.603$ m

   $C_p = 0.707$.

© Springer Nature Singapore Pte Ltd. 2019

B. S. Lee, *Hydrostatics and Stability of Marine Vehicles*, Springer Series on Naval Architecture, Marine Engineering, Shipbuilding and Shipping 7, https://doi.org/10.1007/978-981-13-2682-0

Chapter 3

1. (a)
   Mx = 134.333, My = 175.667
   Centroid is at (5.667, 4.333)
   (b)
   Mx = 41.667, My = 50.833
   Centroid is at (3.389, 2.778)
   (c)
   Mx = 87.625
   My = 98.25
   Centroid is at (1.908, 1.701)
2. Moment about midship = $-22,607.6$ m^3
   Longitudinal centroid = -5.105 m (or 5.105 m aft of midship)
   Longitudinal second moment of area (I_L) = 4776126.9 m^4
   Transverse second moment of area (I_T) = 529498.9 m^4
3. I_L = 1859.18 m^4
   I_T = 3451.67 m^4.

Chapter 4

1. (a) 146.2496 m^2
   (b) 146.2041 m^2
   (c) 145.6896 m^2
2. Difference between Simpson's first and second rules is 0.46%.
   Difference between Simpson's first rule and trapezoidal rule is 1.21%.
3. Area = 5286.12 ft^2
   Centroid is at 19.65 ft aft of midship.
4. ∇ = 2521.32 m^3
   $\Delta = \nabla \times 1.025 = 2459.82$ tonnes
   LCB = -1.582 m from midship
   C_B = 0.708
   C_P = 0.761
   C_M = 0.930
5. ∇ = 2575.2 m^3
   Δ = 2639.58 tonnes
   LCB = -2.261 m from midship
   Waterplane area = 685.92 m^2
   CW = 0.828, CB = 0.723, CM = 0.930, CP = 0.778
   LCF = -1.194 m from midship
   IL = 223941.2 m^4
   IT = 6281.157 m^4.

Chapter 5

1. (a) New draught = 3.075 m
   No, because the ship is wall-sided and therefore LCB does not change for parallel sinkage.
   LCF = -1.277 m I_L = 183777.6 m^4
   (b)
   New mass disp = 2921.3 tonnes
   New volume disp = 2921.3 m^3
   BM_L = 62.91 m
   GM_L = 60.469 m
   MCT = 35.33 tonne-m/cm
   T_{fwd} = 3.045 m
   T_{aft} = 3.183 m
   (c) LCB moves in this case. So, the ship will change trim, albeit rather small.
2. (a) TPC = 32.657 tonnes/m, LCF = -3.286 m
   (b) Vol. disp. = 17411.5 m^3, LCB = -6.842 m
   (c) I_L = 4,600,143 m^4, I_T = 143052.9 m^4
   (d) BM_L = 264.2 m, GM_L = 259.5 m
   (e) T_f = 5.964 m, T_a = 6.150 m. It is assumed that the change in MCT due to the addition of 200 tonnes is negligible.
3. (a) T = 6.585 m
   (b) BM_L = 126.54 m, MCT = 202.223 tonnes-m/cm
   (c) Trimming moment = -31400 tonnes-m, trim = 155.3 cm,
   T_f = 5.809 m, T_a = 7.362 m
4. (a) C_M = 1.0, C_P = 0.8, C_B = 0.8
   (b) TPC = 24.6 tonnes/cm
   (c) LCF = -4.583 m
   (d) I_T = 157,500 m^4, I_L = 1,379,583 m^4
   (e) BM_L = 143.707 m, MCT = 141.407 tonnes-m/cm
   (f) T_f = 3.884 m, T_a = 4.096 m
5. Original draught fwd = 9.127 m
   Amount of ballast discharged = 114.65 tonnes
6. Draught aft = 7.25 m
   Distance to be moved = 36.36 m towards stern.

Chapter 6

1. (a) GM = 0.278 m
   (b) Max KG = 4.278 m
2. GM = 1.556 m
   GM_f = 0.440 m
   GM_f = 0.399 m
3. GM_f = -0.274 m
4. (a) GM = 3.309 m, BM = 6.25 m, KG = 4.441 m
   (b) GM = -0.556 m
   (c) Max allowable cargo on deck 934 tonnes.

5. GM = 1.794 m
   New draught = 25.753 m
   New KG = 19.172 m
   New GM = 1.288 m
6. (a) GM = 1.681 m
   (b) virtual KG = 10.65 m effective GM = 1.030 m
   (c) New GM = 1.474 m
   (d) KG is reduced by about 1 m, but its effect is reduced because BM is reduced
   for higher displacement.
7. (a) KG = 7.391 m
   (b) about 5.5°.

Chapter 7

1. GM = −1.7 cm, 45° (i.e. corner down)
2. about 11°.
3. (a)

Heel angle (°)	5	10	15	18
GZ (m)	0.123	0.264	0.443	0.578

   (b) about 12.82 m.
4. 7.018 m
5. $GZ = 0.7\phi - 0.58\phi^3 + 0.1\cos\phi - 0.05\sin\phi$
   Angle of loll is about 9° to port.

Chapter 8

1. About 21°. Gust lever = 1.5 times steady wind lever.
2. (a) $GZ = 0.7\phi - 0.58\phi^3 + 0.1\phi^5 + 0.05\cos\phi$
   Angle of loll is about 4° to port.
   (b) 12° to starboard
3. (a) 12.759 m
   (b) about 15°
4. $GZ = \left(2.126 + \frac{1}{2}3.626\tan^2\phi\right)\sin\phi$
5. 1.6°
6. (a) wind heel lever = 0.034 m, gust heel lever = 0.051 m
   (b) about 25°.

Index

© Springer Nature Singapore Pte Ltd. 2019
B. S. Lee, *Hydrostatics and Stability of Marine Vehicles*, Springer Series on Naval
Architecture, Marine Engineering, Shipbuilding and Shipping 7,
https://doi.org/10.1007/978-981-13-2682-0

Printed in the United States
By Bookmasters